BEGINNING
ANDROID™ TABLET APPLICATION DEVELOPMENT

Android™ Tablet Application Development

Wei-Meng Lee

Wiley Publishing, Inc.

Beginning Android™ Tablet Application Development

Published by
Wiley Publishing, Inc.
10475 Crosspoint Boulevard
Indianapolis, IN 46256
www.wiley.com

Copyright © 2011 by Wiley Publishing, Inc., Indianapolis, Indiana

Published simultaneously in Canada

ISBN: 978-1-118-10673-0
ISBN: 978-1-118-15075-7 (ebk)
ISBN: 978-1-118-15077-1 (ebk)
ISBN: 978-1-118-15076-4 (ebk)

Manufactured in the United States of America

10 9 8 7 6 5 4 3 2 1

For general information on our other products and services please contact our Customer Care Department within the United States at (877) 762-2974, outside the United States at (317) 572-3993 or fax (317) 572-4002.

Wiley also publishes its books in a variety of electronic formats and by print-on-demand. Not all content that is available in standard print versions of this book may appear or be packaged in all book formats. If you have purchased a version of this book that did not include media that is referenced by or accompanies a standard print version, you may request this media by visiting http://booksupport.wiley.com. For more information about Wiley products, visit us at www.wiley.com.

Library of Congress Control Number: 2011930129

To my family:

Thanks for the understanding and support while I worked on getting this book ready. I love you all!

CREDITS

ABOUT THE AUTHOR

WEI-MENG LEE is a technologist and founder of Developer Learning Solutions (www.learn2develop .net), a technology company specializing in hands-on training on the latest mobile technologies. Wei-Meng has many years of training experience and his training courses place special emphasis on the learning-by-doing approach. This hands-on approach to learning programming makes understanding the subject much easier than reading books, tutorials, and other documentation.

Wei-Meng is also the author of *Beginning iOS 4 Application Development* (Wrox, 2010) and *Beginning Android Application Development* (Wrox, 2011). Contact Wei-Meng at weimenglee@learn2develop.net.

ABOUT THE TECHNICAL EDITOR

KUNAL MITTAL serves as an Executive Director of Technology at Sony Pictures Entertainment where he is responsible for the SOA, Identity Management, and Content Management programs. Kunal is an entrepreneur who helps startups define their technology strategy, product roadmap, and development plans. He generally works in an Advisor or Consulting CTO capacity, and serves actively in the Project Management and Technical Architect functions.

He has authored and edited several books and articles on J2EE, Cloud Computing, and mobile technologies. He holds a Master's degree in Software Engineering and is an instrument-rated private pilot.

ACKNOWLEDGMENTS

WRITING THIS BOOK HAS BEEN A roller-coaster ride. Working with just-released software is always a huge challenge. When I first started work on this book, the Android 3.0 SDK had just been released, and wading through the documentation was like finding a needle in a haystack. To add to the challenge, the Android emulator for the tablet is extremely slow and unstable, making the development process very slow and painful.

Well, now that the book is done, I hope your journey will not be as eventful as mine. Like a good guide, my duty is to make your foray into Android tablet development an enjoyable and fruitful experience. The book you are now holding is the result of the collaborative efforts of many people, and I wish to take this opportunity to acknowledge them here.

First, my personal gratitude to Bob Elliott, executive editor at Wrox. Bob is always ready to lend a listening ear and to offer help when it's needed. It is a great pleasure to work with Bob, as he is one of the most responsive persons I have ever worked with! Thank you, Bob, for the help and guidance!

Of course, I cannot forget Ami Sullivan, my editor (and friend!), who is always a pleasure to work with. After working together on four books, we now know each other so well that we know the content of incoming e-mail messages even before we open them! Thank-you, Ami!

Nor can I forget the heroes behind the scenes: copy editor Luann Rouff and technical editor Kunal Mittal. They have been eagle-eye editing the book, making sure that every sentence makes sense — both grammatically as well as technically. Thanks, Luann and Kunal!

Last, but not least, I want to thank my parents and my wife, Sze Wa, for all the support they have given me. They have selflessly adjusted their schedules to accommodate my busy schedule when I was working on this book. My wife, as always, has stayed up with me on numerous nights as I was furiously working to meet the deadlines, and for this I would like to say to her and my parents: "I love you all!" Finally, to our lovely dog, Ookii, thanks for staying by our side.

CONTENTS

INTRODUCTION

I FIRST STARTED PLAYING WITH THE ANDROID SDK before it was officially released as a 1.0 release. Back then, the tools were unpolished, the APIs in the SDK were unstable, and the documentation was sparse. Fast forward two and a half years, Android is now a formidable mobile operating system, with a following no less impressive that the iPhone. Having gone through all the growing pains of Android, I think now is the best time to start learning about Android programming — the APIs have stabilized and the tools have improved. But one thing remains: Getting started is still an elusive goal for many. What's more, Google has recently released their latest version of the Android SDK — 3.0, for tablet development. The Android 3.0 SDK comes with several new features for tablet developers, and understanding all these new features requires some effort on the part of beginners. It was with this mission in mind that I was motivated to write a book that beginning Android tablet programmers could appreciate, and one that would enable them to write progressively sophisticated applications.

This book was written to help jump-start beginning Android developers, in particular developers targeting tablet devices. It covers just enough for you to get started with tablet programming using Android. You will learn the basics of the new features in Android 3.0. For a more comprehensive overview of the various programming capabilities of Android, I suggest you look at my other book, *Beginning Android Application Development* (Wrox, 2011).

To make the learning interesting, this book walks through the process of building two projects. The first project shows how to build a mapping application for your Android tablet. You will be able to monitor your current location using the built-in GPS, cellular, and wireless network connections. In addition, you will be able to view your location using the Google Maps. The second project demonstrates how to build a pair of location tracker applications, allowing you to track the geographical locations of other Android users through the use of SMS messaging. These two projects serve as a solid starting point for building real-life tablet applications. Have fun!

WHO THIS BOOK IS FOR

This book is for the beginning Android tablet developer who wants to start developing tablet applications using the Google's Android 3.0 SDK. To truly benefit from this book, you should have some background in programming and at least be familiar with object-oriented programming concepts. If you are totally new to Java — the language used for Android development — you might want to take a programming course in Java programming first, or grab one of many good books on Java programming. In my experience, if you already know C# or VB.NET, learning Java is not too much of an effort; you should be comfortable just following along with the *Try It Out* exercises.

For those totally new to programming, I know the lure of developing mobile apps and making some money is simply too tempting to miss. However, I think a better starting point is learning the basics of programming before attempting to try out the examples in this book.

NOTE *All the examples discussed in this book were written and tested using version 2.x and 3.0 of the Android SDK. While every effort has been made to ensure that all the tools used in this book are the latest, it is likely that by the time you read this, a newer version of the tools may be available. As such, some of the instructions/screenshots may differ slightly. However, any changes should be minimal and you should not have any problems following along.*

WHAT THIS BOOK COVERS

This book covers the fundamentals of Android programming using the Android SDK. It is divided into six chapters and three appendices.

Chapter 1: Getting Started with Android Programming for Tablets covers the basics of the Android OS and its current state. You will learn about the features of Android devices, as well as some of the popular devices on the market. You will then learn how to download all the required tools to develop Android applications and then test them on the various types of Android emulators.

Chapter 2: Components of an Android Tablet Application covers the various parts that make up an Android tablet application and some of the new features in Android 3.0 that are specifically designed for tablet applications. In particular, you will learn about the fragment and Action Bar APIs new in Android 3.0, and how you can make use of them to develop compelling tablet applications.

Chapter 3: Android User Interface covers the various components that make up the UI of an Android application. You will learn about the different layouts you can use to build the UI of your application, and the numerous events that are associated with the UI when users interact with the application. You will also learn about the specialized fragments available for Android tablet applications.

Chapter 4: Creating Location-Based Services Applications shows how to make use of Google Maps in your Android application, and how to manipulate it programmatically. In addition, you will learn how to obtain your geographical location using the `LocationManager` class available in the Android SDK. By the end of the chapter, you will have created a very cool Android tablet mapping project.

Chapter 5: SMS Messaging and Networking demonstrates how to send and receive SMS messages programmatically from within your Android application. You will also learn how to use the HTTP protocol to talk to web servers so that you can download text and binary data. The last part of this chapter shows you how to parse XML files to extract the relevant parts of an XML file — a technique that is useful if you are accessing Web services. By the end of this chapter, you will have built a functional location tracker application!

Chapter 6: Publishing Android Applications discusses the various ways you can publish your Android applications when you are ready. You will also learn about the steps to publishing and selling your applications on the Android Market.

Appendix A: Using Eclipse for Android Development provides a quick run-through of the many features in Eclipse.

Appendix B: Using the Android Emulator provides tips and tricks on using the Android emulator to test your applications.

Appendix C: Answers to Exercises contains the solutions to the end-of-chapter exercises found in every chapter.

HOW THIS BOOK IS STRUCTURED

This book breaks down the task of learning Android programming into several smaller chunks, enabling you to digest each topic before delving into a more advanced one.

If you are a total beginner to Android programming, start with Chapter 1. Once you are comfortable with the basics here, head on to the appendices to read more about Eclipse and the Android emulator. When you are ready, you can continue with Chapter 2 and gradually move into more advanced concepts.

A key feature of this book is that all the code samples in each chapter are independent of those discussed in previous chapters. This gives you the flexibility to dive into the topics that interest you most and start working on the *Try It Out* projects.

WHAT YOU NEED TO USE THIS BOOK

All the examples in this book run on the Android emulator (which is included with the Android SDK). However, to get the most out of this book, having a real Android device would be optimal (though not absolutely necessary).

CONVENTIONS

To help you get the most from the text and keep track of what's happening, we've used a number of conventions throughout the book.

TRY IT OUT **These Are Exercises or Examples for You to Follow**

The *Try It Out* exercises appear once or more per chapter as exercises to work through as you follow the text in the book.

1. They usually consist of a set of numbered steps.

2. Follow the steps through with your copy of the project files.

How It Works

After each *Try It Out*, the code you've typed is explained in detail.

As for other conventions in the text:

➤ New terms and important words are *highlighted* in italics when first introduced.

➤ Keyboard combinations are treated like this: Control+R.

➤ Filenames, URLs, and code within the text are treated like so: `persistence.properties`.

Code is presented in two different ways:

```
We use a monofont type with no highlighting for most code examples.
```

**We use bold to emphasize code that is of particular importance in the
present context.**

 WARNING *Boxes like this one hold important, not-to-be forgotten information
that is directly relevant to the surrounding text.*

 NOTE *Notes, tips, hints, tricks, and asides to the current discussion look
like this.*

SOURCE CODE

As you work through the examples in this book, you may choose either to type in all the code
manually or to use the source code files that accompany the book. All the source code used in this
book is available for download at www.wrox.com. When at the site, simply locate the book's title
(use the Search box or one of the title lists) and click the Download Code link on the book's detail
page to obtain all the source code for the book.

 NOTE *Because many books have similar titles, you may find it easiest to search
by ISBN; this book's ISBN is 978-1-118-10673-0.*

Code that is included on the website is highlighted by the following CodeNote:

code snippet filename

After you download the code, just decompress it with your favorite compression tool. Alternatively,
go to the main Wrox code download page at www.wrox.com/dynamic/books/download.aspx to see
the code available for this book and all other Wrox books.

ERRATA

We make every effort to ensure that there are no errors in the text or in the code. However, no one is
perfect, and mistakes do occur. If you find an error in one of our books, such as a spelling mistake or a
faulty piece of code, we would be very grateful for your feedback. By sending in errata, you may save
another reader hours of frustration and at the same time help us provide even higher-quality information.

To find the errata page for this book, go to www.wrox.com and locate the title using the Search box or one of the title lists. Then, on the book details page, click the Book Errata link. On this page, you can view all errata that has been submitted for this book and posted by Wrox editors.

 NOTE *A complete book list, including links to each book's errata, is also available at www.wrox.com/misc-pages/booklist.shtml.*

If you don't spot "your" error on the Book Errata page, go to www.wrox.com/contact/techsupport.shtml and complete the form there to send us the error you have found. We'll check the information and, if appropriate, post a message to the book's errata page and fix the problem in subsequent editions of the book.

P2P.WROX.COM

For author and peer discussion, join the P2P forums at p2p.wrox.com. The forums are a web-based system for you to post messages relating to Wrox books and related technologies and interact with other readers and technology users. The forums offer a subscription feature to e-mail you topics of interest of your choosing when new posts are made to the forums. Wrox authors, editors, other industry experts, and your fellow readers are present on these forums.

At p2p.wrox.com, you will find a number of different forums that will help you not only as you read this book but also as you develop your own applications. To join the forums, just follow these steps:

1. Go to p2p.wrox.com and click the Register link.

2. Read the terms of use and click Agree.

3. Complete the required information to join as well as any optional information you want to provide and click Submit.

4. You will receive an e-mail with information describing how to verify your account and complete the joining process.

 NOTE *You can read messages in the forums without joining P2P, but in order to post your own messages, you must join.*

After you join, you can post new messages and respond to messages that other users post. You can read messages at any time on the Web. If you want to have new messages from a particular forum e-mailed to you, click the Subscribe to This Forum icon by the forum name in the forum listing.

For more information about how to use the Wrox P2P, be sure to read the P2P FAQs for answers to questions about how the forum software works as well as for many common questions specific to P2P and Wrox books. To read the FAQs, click the FAQ link on any P2P page.

PART I
Quick Tour of Android 3 for Tablets

1

Getting Started with Android Programming for Tablets

WHAT YOU WILL LEARN IN THIS CHAPTER

➤ What is Android?

➤ Android versions and its feature set

➤ The Android architecture

➤ The various Android devices on the market

➤ The Android Market application store

➤ How to obtain the tools and SDK for developing Android applications

➤ How to develop your first Android application

Welcome to the world of Android! When I was writing my first book on Android (which was just a couple of months ago), I stated that Android was ranked second in the U.S. smartphone market, second to Research In Motion's (RIM) BlackBerry, and overtaking Apple's iPhone. Shortly after the book went to press, comScore (a global leader in measuring the digital world and the preferred source of digital marketing intelligence) reported that Android has overtaken BlackBerry as the most popular smartphone platform in the U.S.

Indeed. With Google's recent introduction of Android 3.0, code-named *Honeycomb*, it's a perfect time to start learning about Android programming. In my first book, *Beginning Android Application Development* (Wrox, 2011), I focused on getting readers started with the building blocks of Android programming, with particular emphasis on developing applications for Android smartphone applications. With the release of Android 3.0, Google's focus in this new SDK is the introduction of several new features designed for wide-screen devices, specifically tablets. This focus was the impetus behind the book you are currently holding. Therefore, it also focuses on the various features that are specific to wide-screen devices, and contains enough information that can get you jumpstarted with Android tablet development quickly. Readers who want more comprehensive coverage on Android development in general should start with my *Beginning Android Application Development* book first, and then read this book for information on designing for tablets.

In this chapter you will learn what Android is, and what makes it so compelling to both developers and device manufacturers alike. You will also get started with developing your first Android application, and learn how to obtain all the necessary tools and set them up so that you can test your application on an Android 3.0 tablet emulator. By the end of this chapter, you will be equipped with the basic knowledge you need to explore more sophisticated techniques and tricks for developing your next killer Android tablet application.

WHAT IS ANDROID?

Android is a mobile operating system that is based on a modified version of Linux. It was originally developed by a startup of the same name, Android, Inc. In 2005, as part of its strategy to enter the mobile space, Google purchased Android and took over its development work (as well as its development team).

Google wanted Android to be open and free; hence, most of the Android code was released under the open-source Apache License, which means that anyone who wants to use Android can do so by downloading the full Android source code. Moreover, vendors (typically hardware manufacturers) can add their own proprietary extensions to Android and customize Android to differentiate their products from others. This simple development model makes Android very attractive and has thus piqued the interest of many vendors. This has been especially true for companies affected by the phenomenon of Apple's iPhone, a hugely successful product that revolutionized the smartphone industry. Such companies include Motorola and Sony Ericsson, which for many years have been developing their own mobile operating systems. When the iPhone was launched, many of these manufacturers had to scramble to find new ways to revitalize their products. These manufacturers see Android as a solution — they will continue to design their own hardware and use Android as the operating system that powers it.

The main advantage of adopting Android is that it offers a unified approach to application development. Developers need only develop for Android, and their applications should be able to run on numerous different devices, as long as the devices are powered using Android. In the world of smartphones, applications are the most important part of the success chain. Device manufacturers therefore see Android as their best hope to challenge the onslaught of the iPhone, which already commands a large base of applications.

Android Versions

Android has gone through quite a number of updates since its first release. Table 1-1 shows the various versions of Android and their codenames.

TABLE 1-1: A Brief History of Android Versions

ANDROID VERSION	RELEASE DATE	CODENAME
1.1	9 February 2009	
1.5	30 April 2009	Cupcake
1.6	15 September 2009	Donut
2.0/2.1	26 October 2009	Eclair
2.2	20 May 2010	Froyo
2.3	6 December 2010	Gingerbread
3.0	22 February 2011	Honeycomb

In February 2011, Google released Android 3.0, a tablet-only release supporting wide-screen devices. The key changes in Android 3.0 are as follows:

➤ New user interface optimized for tablets

➤ 3D desktop with new widgets

➤ Refined multi-tasking

➤ New web browser features, such as tabbed browsing, form auto-fill, bookmark syncing, and private browsing

➤ Support for multicore processors

Applications written for versions of Android prior to 3.0 are compatible with Android 3.0 devices, and they run without modifications. Android 3.0 tablet applications that make use of the newer features available in 3.0, on the other hand, will not be able to run on older devices. If you want to ensure that an Android tablet application is able to run on all versions of devices, you must programmatically ensure that you only make use of features that are supported in specific versions of Android. To do so, you can make use of the `android.os.Build.VERSION.SDK` constant. The following code snippet shows how you can determine the version of the device during runtime:

```
int version =
    Integer.parseInt(android.os.Build.VERSION.SDK);
switch (version) {
case 8:
    //---use features specific to Android 2.2---
    break;
case 9:
    //---use features specific to Android 2.3.1---
    break;
case 10:
    //---use features specific to Android 2.3.3---
    break;
```

```
case 11:
    //---use features specific to Android 3.0---
    break;
}
```

Android Devices in the Market

Android devices come in all shapes and sizes. As of late May 2010, the Android OS powers all of the following types of devices:

➤ Smartphones

➤ Tablets

➤ E-reader devices

➤ Netbooks

➤ MP4 players

➤ Internet TVs

Increasingly, manufacturers are rushing out to release Android tablets. Tablet sizes typically start at seven inches, measured diagonally. Figure 1-1 shows the Samsung Galaxy Tab (top), a seven-inch tablet, and the Dell Streak (bottom), a five-inch tablet.

While the Samsung Galaxy Tab and the Dell Streak run the older Android 2.x, the newer tablets run the latest Android 3.0 Honeycomb. Figure 1-2 shows the Motorola Xoom.

FIGURE 1-1

FIGURE 1-2

Besides the Motorola Xoom, the LG Optimus
Pad, shown in Figure 1-3, is another Android 3.0
device, running the latest Android Honeycomb OS.

The Android Market

As mentioned earlier, one of the main factors
determining the success of a smartphone platform
is the applications that support it. It is clear from
the success of the iPhone that applications play
a very vital role in determining whether a new
platform swims or sinks. In addition, making
these applications accessible to the general user is
extremely important.

FIGURE 1-3

As such, in August 2008, Google announced the
Android Market, an online application store for Android devices, and made it available to users in
October 2008. Using the Market application that is preinstalled on their Android device, users can
simply download third-party applications directly onto their devices. Both paid and free applications
are supported on the Android Market, though paid applications are available only to users in certain
countries due to legal issues.

Similarly, in some countries, users can buy paid applications from the Android Market, but
developers cannot sell in that country. As an example, at the time of writing, users in India can buy
apps from the Android Market, but developers in India cannot sell apps on the Android Market.
The reverse may also be true; for example, users in South Korea cannot buy apps on the Android
Market, but developers in South Korea can sell apps on it.

OBTAINING THE REQUIRED TOOLS

Naturally, you are anxious to get your hands dirty and start writing some applications! Before you
write your first tablet application, however, you need to download the required tools and SDKs.

For Android development, you can use a Mac, a Windows PC, or a Linux machine. All the tools
needed are free and can be downloaded from the Web. All the examples provided in this book will
work fine with the Android emulator.

> **NOTE** *This book uses a Windows 7 computer to demonstrate all the code samples.
> If you are using a Mac or a Linux computer, the screenshots should look similar; minor
> differences may be present, but you should be able to follow along without problems.*

So, let the fun begin!

Java JDK

The Android SDK makes use of the Java SE Development Kit (JDK). Hence, if your computer does
not have the JDK installed, you should start off by downloading the JDK from www.oracle.com/

`technetwork/java/javase/downloads/index.html` and installing it prior to moving to the next section.

Eclipse

The first step toward developing any applications is obtaining the *integrated development environment (IDE).* In the case of Android, the recommended IDE is Eclipse, a multi-language software development environment featuring an extensible plug-in system. It can be used to develop various types of applications, using languages such as Java, Ada, C, C++, COBOL, Python, and others.

For Android development, you should download the Eclipse IDE for Java EE Developers (`www .eclipse.org/downloads/packages/eclipse-ide-java-ee-developers/heliossr1`). Six editions are available: Windows (32 and 64-bit), Mac OS X (Cocoa 32 and 64), and Linux (32 and 64-bit). Simply select the relevant one for your operating system. All the examples in this book were tested using the 32-bit version of Eclipse for Windows.

Once the Eclipse IDE is downloaded, unzip its contents (the `eclipse` folder) into a folder, say `C:\Android\`.

Downloading the Android SDK

The next important piece of software you need to download is, of course, the Android SDK. The Android SDK contains a debugger, libraries, an emulator, documentation, sample code, and tutorials.

You can download the Android SDK from `http://developer.android.com/sdk/index.html` (see Figure 1-4).

FIGURE 1-4

For Windows users, there are two ways in which you can download the Android SDK — either you download the entire Android SDK package — `android-sdk_r10-windows.zip` or you can download the SDK installer — `installer_r10-windows.zip`. For beginning Android developers, I strongly encourage you to download the latter, as it makes it very easy for you to get started.

Once the `installer_r10-windows.zip` package is downloaded, double-click on it to start the installation process. It will first detect whether the JDK is installed and will only continue if it finds one installed on your computer. Next, you will be asked to choose a destination folder for installing the SDK (see Figure 1-5). Remember the path to this folder because you need to use it later.

Click Next to continue.

You will next be asked to choose a Start Menu folder to install the Android SDK shortcut. Use the default Android SDK Tools folder and click Install. When the installation is complete, click Finish (see Figure 1-6). Doing so will start the SDK Manager, which downloads all the necessary packages for you to test your Android applications.

FIGURE 1-5

FIGURE 1-6

Installing the Packages

When the SDK Manager is started, it first checks for the packages that are available for installation. The packages contain the documentation and SDK specific to each version of the Android OS. They also contain sample code and tools for the various platforms.

Figure 1-7 shows the various SDK packages that you can install on your computer. Double-click on each package name to select or deselect a package. If you are not sure which packages to install, you might want to select the Accept All radio button to download and install all the packages.

Click Install to proceed with the downloading and installation of the various selected packages.

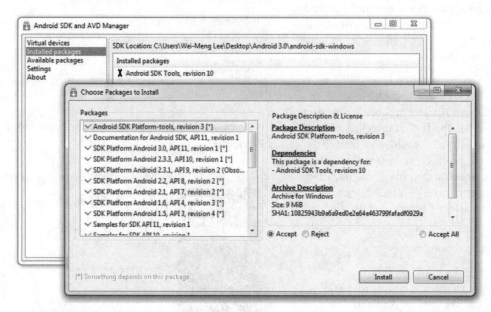

FIGURE 1-7

Each version of the Android OS is identified by an API level number. For example, Android 3.0 is level 11 (API 11), while Android 2.3.3 is level 10 (API 10), and so on. For each level, two platforms are available. For example, level 11 offers the following:

➤ SDK Platform Android 3.0

➤ Google APIs by Google Inc., Android API 11, revision 1

The key difference between the two is that the Google APIs platform contains the Google Maps library. Therefore, if the application you are writing requires Google Maps, you need to create an AVD using the Google APIs platform.

Downloading and installing the packages takes some time, so you have to be patient. When all the packages are installed, click Close. You should now see a listing of all the packages installed (see Figure 1-8).

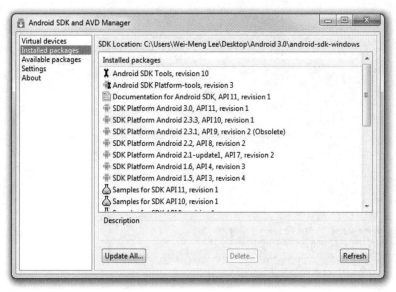

FIGURE 1-8

Creating Android Virtual Devices (AVDs)

Once the packages are downloaded and installed, the next step is to create an Android Virtual Device (AVD) to be used for testing your Android applications. An AVD is an emulator instance that enables you to model an actual device. Each AVD consists of a hardware profile, a mapping to a system image, as well as emulated storage, such as a secure digital (SD) card.

You can create as many AVDs as you want in order to test your applications with several different configurations. This testing is important to confirm that your application behaves as expected when it is run on different devices with varying capabilities.

 NOTE *Appendix B discusses some of the capabilities of the Android emulator.*

To create an AVD, select the Virtual Devices item in the left pane of the Android SDK and AVD Manager window (see Figure 1-9).

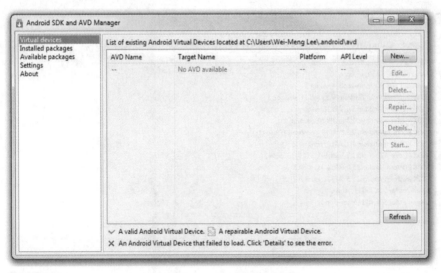

FIGURE 1-9

Then click the New… button located in the right pane of the window. In the Create new Android Virtual Device (AVD) window, enter the items as shown in Figure 1-10. Click the Create AVD button when you are done.

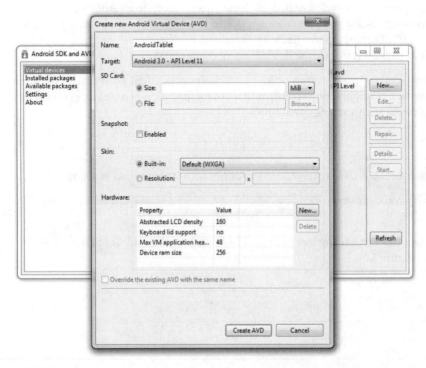

FIGURE 1-10

In this case, you have created an AVD (put simply, an Android emulator) that emulates an Android device running version 3.0 of the OS. In addition to what you have created, you also have the option to emulate the device with an SD card and different screen densities and resolutions.

 NOTE *Appendix B explains how to emulate the different types of Android devices.*

It is preferable to create a few AVDs with different API levels so that your application can be tested on different devices. To emulate the Motorola Xoom, you should choose the "Google APIs (Google Inc.) – API Level 11" target.

To see what the Android emulator looks like, select the AVD you have just created and click the Start... button. Figure 1-11 shows the Android 3.0 emulator.

FIGURE 1-11

Click and move the lock icon to touch a circle that appears when you move the mouse. This unlocks the emulator. Figure 1-12 shows the main window of the Android 3.0 screen.

FIGURE 1-12

Clicking the Apps icon on the top-right corner of the screen reveals a list of installed applications on the device (see Figure 1-13).

FIGURE 1-13

Android Development Tools (ADT)

With the Android SDK and AVD set up, it is now time to configure Eclipse to recognize the Android project template. The Android Development Tools (ADT) plug-in for Eclipse is an extension to the

Eclipse IDE that supports the creation and debugging of Android applications. Using the ADT, you will be able to do the following in Eclipse:

➤ Create new Android application projects

➤ Access the tools for accessing your Android emulators and devices

➤ Compile and debug Android applications

➤ Export Android applications into Android Packages (APKs)

➤ Create digital certificates for code-signing your APK

To install the ADT, first launch Eclipse by double-clicking the `eclipse.exe` file located in the `eclipse` folder.

When Eclipse is first started, you are prompted for a folder to use as your workspace. In Eclipse, a workspace is a folder where you store all your projects. Take the default suggestion and click OK.

Once Eclipse is up and running, select the Help ⇨ Install New Software… menu item (see Figure 1-14).

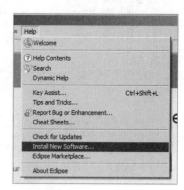

FIGURE 1-14

In the Install window that appears, type **http://dl-ssl.google.com/android/eclipse** in the topmost text box (see Figure 1-15) and press Enter.

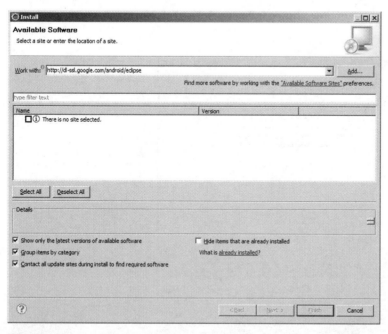

FIGURE 1-15

After a while, you will see the Developer Tools item appear in the middle of the window (see Figure 1-16). Expand it and it will reveal its contents: Android DDMS, Android Development Tools, Android Hierarchy Viewer, and Android Traceview. Check all of them and click Next.

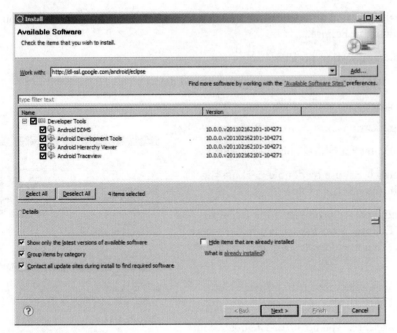

FIGURE 1-16

When you see the Install Details window, shown in Figure 1-17, click Next.

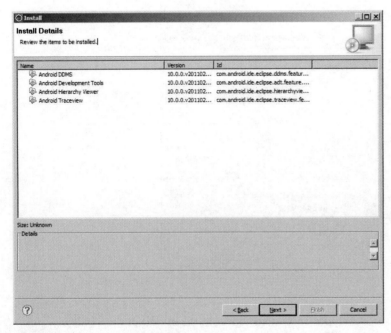

FIGURE 1-17

You will be asked to review the licenses for the tools. Check the option to accept the license agreements (see Figure 1-18). Click Finish to continue.

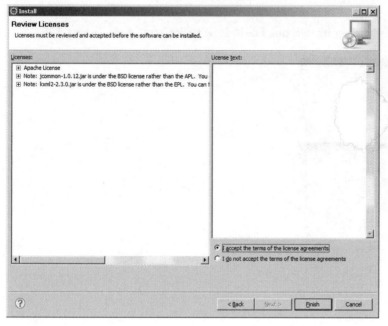

FIGURE 1-18

Eclipse proceeds to download the tools from the Internet and install them. This takes some time, so be patient.

 NOTE *If you have any problems downloading the ADT, check out Google's help at* `http://developer.android.com/sdk/eclipse-adt.html#installing`.

Once the ADT is installed, you will be prompted to restart Eclipse. After doing so, select Window ⇨ Preferences (see Figure 1-19).

In the Preferences window that appears, select Android. Enter the location of the Android SDK folder (that you supplied earlier when you downloaded and installed the Android SDK). Click OK.

CREATING YOUR FIRST ANDROID APPLICATION

With all the tools and the SDK downloaded and installed, it is now time to start your engine! As in all programming books, the first example uses the ubiquitous Hello World application. This will enable you to have a detailed look at the various components that make up an Android project.

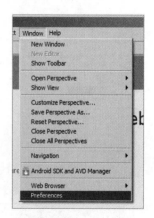

FIGURE 1-19

TRY IT OUT	Creating Your First Android Application

codefile HelloWorld.zip available for download at Wrox.com

1. Using Eclipse, create a new project by selecting File ⇨ New ⇨ Project... (see Figure 1-20).

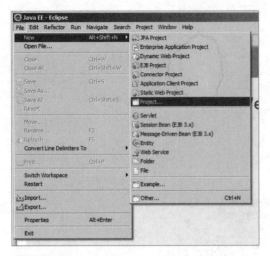

FIGURE 1-20

NOTE *After you have created your first Android application, subsequent Android projects can be created by selecting File ⇨ New ⇨ Android Project.*

2. Expand the Android folder and select Android Project (see Figure 1-21).

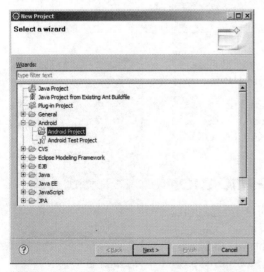

FIGURE 1-21

3. Name the Android project as shown in Figure 1-22 and then click Finish.

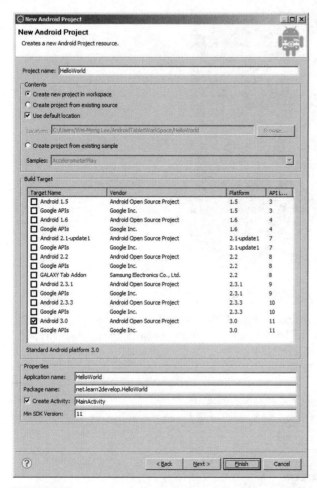

FIGURE 1-22

 NOTE *You need to have at least a period (.) in the package name. The recommended convention for the package name is to use your domain name in reverse order, followed by the project name. For example, my company's domain name is* `learn2develop.net`, *hence my package name would be* `net` `.learn2develop.HelloWorld`.

4. The Eclipse IDE should now look like Figure 1-23.

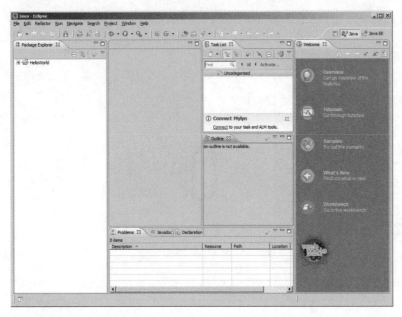

FIGURE 1-23

5. In the Package Explorer (located on the left of the Eclipse IDE), expand the `HelloWorld` project by clicking the various arrows displayed to the left of each item in the project. In the `res/layout` folder, double-click the `main.xml` file (see Figure 1-24).

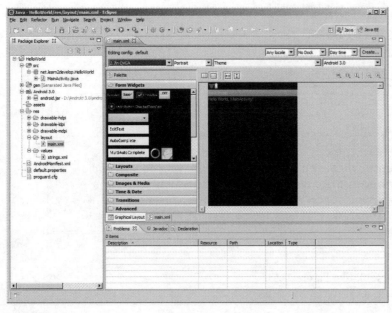

FIGURE 1-24

6. The `main.xml` file defines the user interface (UI) of your application. The default view is the Layout view, which lays out the activity graphically. To modify the UI, click the `main.xml` tab located at the bottom.

7. Add the following code in bold to the `main.xml` file:

```xml
<?xml version="1.0" encoding="utf-8"?>
<LinearLayout xmlns:android="http://schemas.android.com/apk/res/android"
    android:orientation="vertical"
    android:layout_width="fill_parent"
    android:layout_height="fill_parent" >

<TextView
    android:layout_width="fill_parent"
    android:layout_height="wrap_content"
    android:text="@string/hello" />

<TextView
    android:layout_width="fill_parent"
    android:layout_height="wrap_content"
    android:text="This is my first Android Application!" />

<Button
    android:layout_width="fill_parent"
    android:layout_height="wrap_content"
    android:text="And this is a clickable button!" />

</LinearLayout>
```

FIGURE 1-25

8. To save the changes made to your project, press Ctrl+s.

9. You are now ready to test your application on the Android emulator. Select the project name in Eclipse and press F11. You will be asked to select a way to debug the application. Select Android Application as shown in Figure 1-25 and click OK.

> **NOTE** *Some Eclipse installations have an irritating bug: After creating a new project, Eclipse reports that it contains errors when you try to debug the application. This happens even when you have not modified any files or folders in the project. To solve this problem, simply delete the* `R.java` *file located under the* `gen/net.learn2develop.HelloWorld` *folder; Eclipse will automatically generate a new* `R.java` *file for you. Once this is done, the project shouldn't contain any errors.*

10. The Android emulator will now be started (if the emulator is locked, you need to slide the unlock button to unlock it first). Figure 1-26 shows the application running on the Android emulator.

FIGURE 1-26

11. Click the Home button (the house icon in the lower-left corner above the keyboard) so that it now shows the Home screen (see Figure 1-27).

FIGURE 1-27

12. Click the Apps icon to display the list of applications installed on the device. Note that the HelloWorld application is now installed in the application launcher (see Figure 1-28).

FIGURE 1-28

How It Works

To create an Android project using Eclipse, you need to supply the information shown in Table 1-2.

TABLE 1-2: Project Files Created By Default

PROPERTIES	DESCRIPTION
Project name	The name of the project.
Application name	A user-friendly name for your application.
Package name	The name of the package. You should use a reverse domain name for this.
Create Activity	The name of the first activity in your application.
Min SDK Version	The minimum version of the SDK that your project is targeting.

In Android, an activity is a window that contains the user interface of your applications. An application can have zero or more activities; in this example, the application contains one activity: `MainActivity`. This `MainActivity` is the entry point of the application, which is displayed when the application is started. Chapter 2 discusses activities in more detail.

In this simple example, you modified the `main.xml` file to display the string "This is my first Android Application!" and a button. The `main.xml` file contains the user interface of the activity, which is displayed when `MainActivity` is loaded.

When you debug the application on the Android emulator, the application is automatically installed on the emulator. And that's it — you have developed your first Android tablet application!

The next section unravels how all the various files in your Android project work together to make your application come alive.

ANATOMY OF AN ANDROID APPLICATION

Now that you have created your first Hello World Android application, it is time to dissect the innards of the Android project and examine all the parts that make everything work.

FIGURE 1-29

First, note the various files that make up an Android project in the Package Explorer in Eclipse (see Figure 1-29).

The various folders and their files are as follows:

➤ `src` — Contains the .java source files for your project. In this example, there is one file, `MainActivity.java`. The `MainActivity.java` file is the source file for your activity. You will write the code for your application in this file.

➤ `Android 3.0` library — This item contains one file, `android.jar`, which contains all the class libraries needed for an Android application.

➤ `gen` — Contains the `R.java` file, a compiler-generated file that references all the resources found in your project. You should not modify this file.

➤ `assets` — This folder contains all the assets used by your application, such as HTML, text files, databases, etc.

➤ `res` — This folder contains all the resources used in your application. It also contains a few other subfolders: `drawable-<resolution>`, `layout`, and `values`.

➤ `AndroidManifest.xml` — This is the manifest file for your Android application. Here you specify the permissions needed by your application, as well as other features (such as intent-filters, receivers, etc.).

The `main.xml` file defines the user interface for your activity. Observe the following in bold:

```
<TextView
    android:layout_width="fill_parent"
    android:layout_height="wrap_content"
    android:text="@string/hello" />
```

The `@string` in this case refers to the `strings.xml` file located in the `res/values` folder. Hence, `@string/hello` refers to the `hello` string defined in the `strings.xml` file, which is "Hello World, MainActivity!":

```
<?xml version="1.0" encoding="utf-8"?>
<resources>
    <string name="hello">Hello World, MainActivity!</string>
    <string name="app_name">HelloWorld</string>
</resources>
```

It is recommended that you store all the string constants in your application in this `strings.xml` file and reference these strings using the `@string` identifier. That way, if you ever need to localize your application to another language, all you need to do is replace the strings stored in the `strings.xml` file with the targeted language and recompile your application.

Observe the content of the `AndroidManifest.xml` file:

```
<?xml version="1.0" encoding="utf-8"?>
<manifest xmlns:android="http://schemas.android.com/apk/res/android"
    package="net.learn2develop.HelloWorld"
    android:versionCode="1"
    android:versionName="1.0">
  <uses-sdk android:minSdkVersion="11" />
  <application android:icon="@drawable/icon"
          android:label="@string/app_name">
      <activity android:name=".MainActivity"
              android:label="@string/app_name">
          <intent-filter>
              <action android:name="android.intent.action.MAIN" />
              <category android:name="android.intent.category.LAUNCHER" />
          </intent-filter>
      </activity>
  </application>
</manifest>
```

The `AndroidManifest.xml` file contains detailed information about the application:

➤ It defines the package name of the application as `net.learn2develop.HelloWorld`.

➤ The version code of the application is 1. This value is used to identify the version number of your application. It can be used to programmatically determine whether an application needs to be upgraded.

➤ The version name of the application is 1.0. This string value is mainly used for display to the user. You should use the format: *<major>.<minor>.<point>* for this value.

➤ The application uses the image named `icon.png` located in the `drawable` folder.

➤ The name of this application is the string named `app_name` defined in the `strings.xml` file.

➤ There is one activity in the application, represented by the `MainActivity.java` file. The label displayed for this activity is the same as the application name.

➤ Within the definition for this activity, there is an element named `<intent-filter>`:

➤ The action for the intent filter is named `android.intent.action.MAIN` to indicate that this activity serves as the entry point for the application.

➤ The category for the intent filter is named `android.intent.category.LAUNCHER` to indicate that the application can be launched from the device's Launcher icon. Chapter 2 discusses intents in more details.

➤ Finally, the `android:minSdkVersion` attribute of the `<uses-sdk>` element specifies the minimum version of the OS on which the application will run.

As you add more files and folders to your project, Eclipse automatically generates the content of `R.java`, which at the moment contains the following:

```
package net.learn2develop.HelloWorld;

public final class R {
    public static final class attr {
    }
    public static final class drawable {
        public static final int icon=0x7f020000;
    }
    public static final class layout {
        public static final int main=0x7f030000;
    }
    public static final class string {
        public static final int app_name=0x7f040001;
        public static final int hello=0x7f040000;
    }
}
```

You are not supposed to modify the content of the `R.java` file; Eclipse automatically generates the content for you when you modify your project.

 NOTE *If you delete* `R.java` *manually, Eclipse regenerates it for you immediately. Note that in order for Eclipse to generate the* `R.java` *file for you, the project must not contain any errors. If you realize that Eclipse has not regenerated* `R.java` *after you have deleted it, check your project again. The code may contain syntax errors, or your XML files (such as* `AndroidManifest.xml`, `main.xml`, *etc.) may not be well formed.*

Finally, the code that connects the activity to the UI (main.xml) is the setContentView() method, which is in the MainActivity.java file:

```
package net.learn2develop.HelloWorld;

import android.app.Activity;
import android.os.Bundle;

public class MainActivity extends Activity {
    /** Called when the activity is first created. */
    @Override
    public void onCreate(Bundle savedInstanceState) {
        super.onCreate(savedInstanceState);
        setContentView(R.layout.main);
    }
}
```

Here, R.layout.main refers to the main.xml file located in the res/layout folder. As you add additional XML files to the res/layout folder, the filenames are automatically generated in the R.java file. The onCreate() method is one of many methods that are fired when an activity is loaded. Chapter 2 discusses the life cycle of an activity in more detail.

SUMMARY

This chapter provided a brief overview of Android, and highlighted some of its capabilities. If you have followed the sections on downloading the tools and SDK, you should now have a working system — one that is capable of developing more interesting Android applications than the Hello World application. In the next chapter, you will learn about activities and some of the new features in Android 3.0.

EXERCISES

1. What is an AVD?

2. What is the difference between the android:versionCode and android:versionName attributes in the AndroidManifest.xml file?

3. What is the purpose of the strings.xml file?

Answers to the Exercises can be found in Appendix C.

▶ **WHAT YOU LEARNED IN THIS CHAPTER**

TOPIC	KEY CONCEPTS
Android OS	Android is an open-source mobile operating system based on the Linux operating system. It is available to anyone who wants to adapt it to run on their own devices.
Languages used for Android application development	You use the Java programming language to develop Android applications.
Android Market	The Android Market hosts all the various Android applications written by third-party developers.
Tools for Android application development	Eclipse IDE, Android SDK, and the ADT
Activity	An activity is represented by a screen in your Android application. Each application can have zero or more activities.
The Android manifest file	The `AndroidManifest.xml` file contains detailed configuration information for your application. As your application becomes more sophisticated as you progress through the chapters, you modify this file and learn about the different information you can add to it.

2

Components of an Android Tablet Application

WHAT YOU WILL LEARN IN THIS CHAPTER

➤ What are activities in Android?

➤ The new Fragments feature in Android 3.0 for Tablets

➤ The new Action Bar in Android 3.0 applications

In the previous chapter, you learned how to obtain the latest version of the Android SDK, how to start developing your first Android tablet application using Eclipse, and how to test it on the Android emulator. In this chapter, you will learn the various components that make up an Android tablet application and some of the new features in Android 3.0 that are specifically designed for tablet applications. In particular, you will learn about the new fragment and Action Bar APIs in Android 3.0, and how you can use them to develop compelling tablet applications.

ACTIVITIES

In Android, an *activity* is a window that contains the user interface for your application, and users interact directly with the activities of your applications.

 NOTE *If you are new to Android programming, I suggest you read my book* Beginning Android Application Development *(also from Wrox, 2011) to get acquainted with the basic concepts of activities and intents.*

To create an activity, you create a Java class that extends the `Activity` base class:

```
package net.learn2develop.Activities;

import android.app.Activity;
import android.os.Bundle;

public class MainActivity extends Activity {
    /** Called when the activity is first created. */
    @Override
    public void onCreate(Bundle savedInstanceState) {
        super.onCreate(savedInstanceState);
        setContentView(R.layout.main);
    }
}
```

Your activity class would then load its UI component using the XML file defined in your res/ `layout` folder of the project. In this example, you would load the UI from the `main.xml` file:

```
        setContentView(R.layout.main);
```

Every activity you have in your application must be declared in your `AndroidManifest.xml` file, like this:

```
<?xml version="1.0" encoding="utf-8"?>
<manifest xmlns:android="http://schemas.android.com/apk/res/android"
        package="net.learn2develop.Activities"
        android:versionCode="1"
        android:versionName="1.0">
    <application android:icon="@drawable/icon"
        android:label="@string/app_name">
        <activity android:name=".MainActivity"
                android:label="@string/app_name">
            <intent-filter>
                <action android:name="android.intent.action.MAIN" />
                <category
                    android:name="android.intent.category.LAUNCHER" />
            </intent-filter>
        </activity>
    </application>
    <uses-sdk android:minSdkVersion="11" />
</manifest>
```

The `Activity` base class defines a series of events that governs the life cycle of an activity. The `Activity` class defines the following events:

➤ `onCreate()` — Called when the activity is first created

➤ `onStart()` — Called when the activity becomes visible to the user

➤ `onResume()` — Called when the activity starts interacting with the user

➤ `onPause()` — Called when the current activity is being paused and the previous activity is being resumed

➤ onStop() — Called when the activity is no longer visible to the user

➤ onDestroy() — Called before the activity is destroyed by the system (either manually or by the system) to conserve memory

➤ onRestart() — Called when the activity has been stopped and is restarting again

By default, the activity created for you contains the onCreate() event. Within this event handler is the code that helps to display the UI elements of your screen.

Figure 2-1 shows the life cycle of an activity and the various stages it goes through — from when the activity is started until it ends.

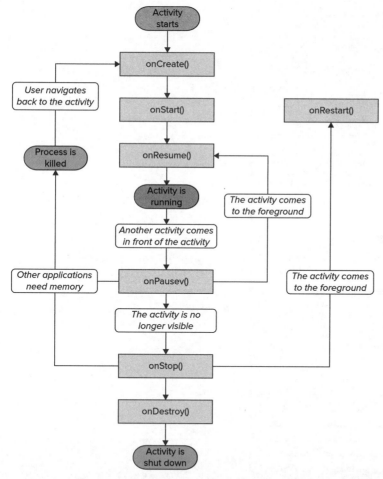

FIGURE 2-1

The best way to understand the various stages of an activity is to create a new project, implement the various events, and then subject the activity to various user interactions.

TRY IT OUT Exploring the Life Cycle of an Activity

codefile Activities.zip available for download at Wrox.com

1. Using Eclipse, create a new Android project and name it as shown in Figure 2-2.

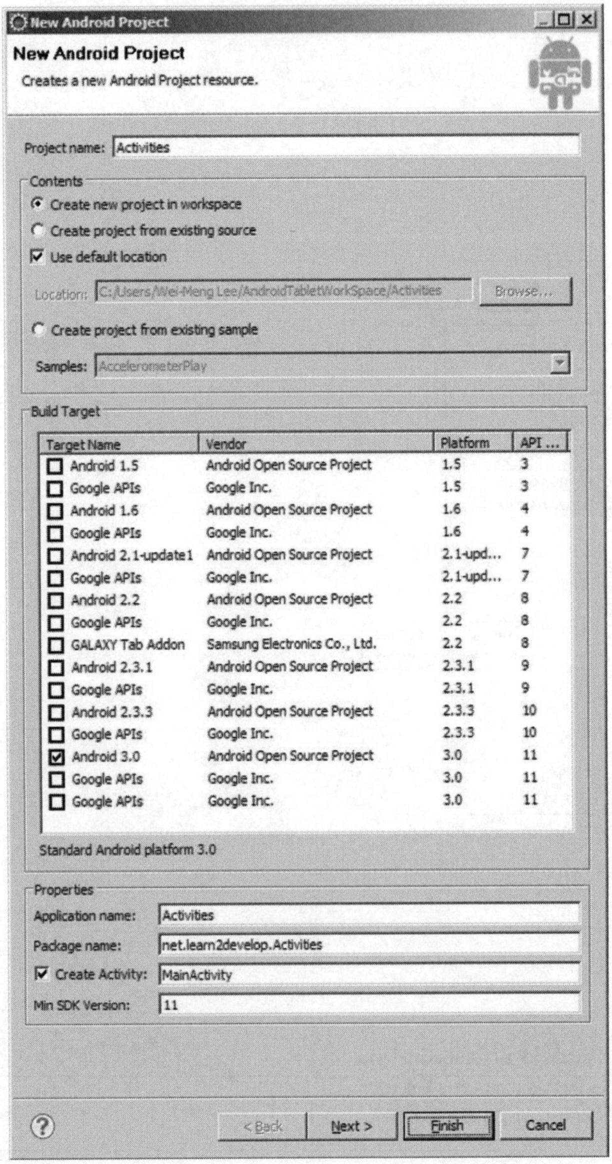

FIGURE 2-2

2. In the `MainActivity.java` file, add the following statements in bold:

```java
package net.learn2develop.Activities;

import android.app.Activity;
import android.os.Bundle;

import android.util.Log;

public class MainActivity extends Activity {
    String tag = "Events";

    /** Called when the activity is first created. */
    @Override
    public void onCreate(Bundle savedInstanceState) {
        super.onCreate(savedInstanceState);
        setContentView(R.layout.main);
        Log.d(tag, "In the onCreate() event");
    }
    public void onStart()
    {
        super.onStart();
        Log.d(tag, "In the onStart() event");
    }
    public void onRestart()
    {
        super.onRestart();
        Log.d(tag, "In the onRestart() event");
    }
    public void onResume()
    {
        super.onResume();
        Log.d(tag, "In the onResume() event");
    }
    public void onPause()
    {
        super.onPause();
        Log.d(tag, "In the onPause() event");
    }
    public void onStop()
    {
        super.onStop();
        Log.d(tag, "In the onStop() event");
    }
    public void onDestroy()
    {
        super.onDestroy();
        Log.d(tag, "In the onDestroy() event");
    }
}
```

3. Press F11 to debug the application on the Android emulator. The activity is shown in Figure 2-3.

FIGURE 2-3

4. When the activity is first loaded, you should see the following in the LogCat window (Window ⇨ Show View ⇨ LogCat; see also Figure 2-4):

```
04-03 01:05:01.256: DEBUG/Events(1222): In the onCreate() event
04-03 01:05:01.256: DEBUG/Events(1222): In the onStart() event
04-03 01:05:01.276: DEBUG/Events(1222): In the onResume() event
```

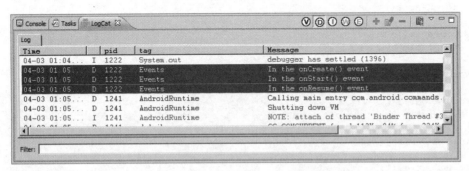

FIGURE 2-4

5. When you now click the Back button on the Android emulator, observe that the following is printed:

```
04-03 01:07:15.785: DEBUG/Events(1222): In the onPause() event
04-03 01:07:17.335: DEBUG/Events(1222): In the onStop() event
04-03 01:07:17.335: DEBUG/Events(1222): In the onDestroy() event
```

6. Click the Recent Apps button (located on the System Bar at the bottom of the screen) and then click the Activities icon (see Figure 2-5). Observe the following printed in the LogCat window:

FIGURE 2-5

```
04-03 01:08:42.446: DEBUG/Events(1222): In the onCreate() event
04-03 01:08:42.446: DEBUG/Events(1222): In the onStart() event
04-03 01:08:42.466: DEBUG/Events(1222): In the onResume() event
```

7. Click the Home button on the Android emulator so that the activity is pushed to the background. Observe the output in the LogCat window:

```
04-03 01:12:54.945: DEBUG/Events(1222): In the onPause() event
04-03 01:12:57.106: DEBUG/Events(1222): In the onStop() event
```

8. Notice that the onDestroy() event is not called, indicating that the activity is still in memory. Click the Apps button and launch the Activities application once more. The activity is now visible again. Observe the output in the LogCat window:

```
04-03 01:18:06.855: DEBUG/Events(1222): In the onRestart() event
04-03 01:18:06.855: DEBUG/Events(1222): In the onStart() event
04-03 01:18:06.865: DEBUG/Events(1222): In the onResume() event
```

The onRestart() event is now fired, followed by the onStart() and onResume() events.

How It Works

As you can see from this simple experiment, an activity is *destroyed* when you press the Back button. This is crucial to know, as whatever state the activity is currently in will be lost; hence, you need to write additional code in your activity to preserve its state when it is destroyed. At this point, note that the `onPause()` event is called in both scenarios — when an activity is sent to the background, as well as when it is killed when the user presses the Back button.

When an activity is started, the `onStart()` and `onResume()` events are always called, regardless of whether the activity is restored from the background or newly created.

 NOTE *Even if an application has only one activity and the activity is killed, the application will still be running in memory.*

FRAGMENTS

The previous section showed you what an activity is. In a small-screen device (such as a smartphone), an activity typically fills up an entire screen, displaying the various views that make up the user interface of an application. The activity is essentially a container for views. However, when an activity is displayed in a large-screen device, such as on a tablet, it is somewhat out of place. Suddenly, the screen becomes much bigger and all the views in an activity must be arranged to make full use of the bigger screen, resulting in complex changes to the view hierarchy. A better approach would be to have "mini-activities," each containing its own set of views. During runtime, an activity can contain one or more of these "mini-activities," depending on the screen orientation in which the device is held. In Android 3.0, these "mini-activities" are known as *fragments*.

Think of a fragment as another form of activity. You create fragments to contain views, just like activities. Fragments are always embedded in an activity. A good way to imagine a fragment is to look at Figure 2-6. Here, you have two fragments. Fragment 1 may contain a `ListView` showing a list of book titles. Fragment 2 may contain some `TextViews` and `ImageViews` showing some text and images.

Now, imagine the application is running on an Android tablet in portrait mode (or on an Android smartphone). In this case, Fragment 1 may be embedded in one activity, while Fragment 2 may be embedded in another activity (see Figure 2-7). When users select an item in the list in Fragment 1, Activity 2 will be started.

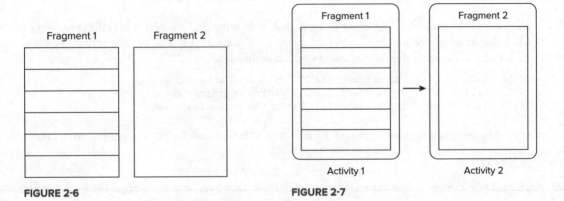

FIGURE 2-6

FIGURE 2-7

If the application is now displayed in a tablet in landscape mode, then both fragments can be embedded within a single activity, as shown in Figure 2-8.

From this discussion, it becomes apparent that fragments present a versatile way in which you can create the user interface of an Android application. Fragments form the atomic unit of your user interface, and they can be dynamically added (or removed) to activities in order to create the best user experience possible for the target device.

The following Try It Out shows you the basics of fragments.

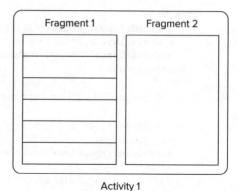

Activity 1

FIGURE 2-8

TRY IT OUT Using Fragments

codefile Fragments.zip available for download at Wrox.com

1. Using Eclipse, create a new Android project and name it **Fragments**.

2. In the res/layout folder, add a new file and name it **fragment1.xml**. Populate it with the following:

```xml
<?xml version="1.0" encoding="utf-8"?>
<LinearLayout
    xmlns:android="http://schemas.android.com/apk/res/android"
    android:orientation="vertical"
    android:layout_width="fill_parent"
    android:layout_height="fill_parent"
    android:background="#00FF00"
    >
<TextView
    android:layout_width="fill_parent"
    android:layout_height="wrap_content"
    android:text="This is fragment #1" />
</LinearLayout>
```

3. Also in the res/layout folder, add another new file and name it **fragment2.xml**. Populate it as follows:

```xml
<?xml version="1.0" encoding="utf-8"?>
<LinearLayout
    xmlns:android="http://schemas.android.com/apk/res/android"
    android:orientation="vertical"
    android:layout_width="fill_parent"
    android:layout_height="fill_parent"
    android:background="#FFFE00"
    >
```

```
<TextView
    android:layout_width="fill_parent"
    android:layout_height="wrap_content"
    android:text="This is fragment #2" />
</LinearLayout>
```

4. In `main.xml`, add the following code in bold:

```
<?xml version="1.0" encoding="utf-8"?>
<LinearLayout xmlns:android="http://schemas.android.com/apk/res/android"
    android:orientation="horizontal"
    android:layout_width="match_parent"
    android:layout_height="match_parent">
    <fragment
        android:name="net.learn2develop.Fragments.Fragment1"
        android:id="@+id/fragment1"
        android:layout_weight="1"
        android:layout_width="0px"
        android:layout_height="match_parent" />
    <fragment
        android:name="net.learn2develop. Fragments.Fragment2"
        android:id="@+id/fragment2"
        android:layout_weight="1"
        android:layout_width="0px"
        android:layout_height="match_parent" />
</LinearLayout>
```

5. Under the `net.learn2develop.Fragments` package name, add two Java class files and name them **Fragment1.java** and **Fragment2.java** (see Figure 2-9).

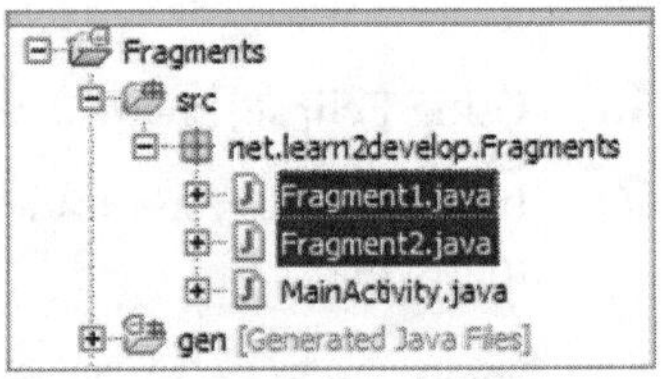

FIGURE 2-9

6. Add the following code to `Fragment1.java`:

```
package net.learn2develop.Fragments;

import net.learn2develop.Fragments.R;
import android.app.Fragment;
import android.os.Bundle;
import android.view.LayoutInflater;
import android.view.View;
import android.view.ViewGroup;

public class Fragment1 extends Fragment {
    @Override
    public View onCreateView(LayoutInflater inflater,
    ViewGroup container, Bundle savedInstanceState) {
        // Inflate the layout for this fragment
        return inflater.inflate(
            R.layout.fragment1, container, false);
    }
}
```

7. Add the following code to `Fragment2.java`:

```
package net.learn2develop.Fragments;

import net.learn2develop.Fragments.R;
import android.app.Fragment;
import android.os.Bundle;
import android.view.LayoutInflater;
import android.view.View;
import android.view.ViewGroup;

public class Fragment2 extends Fragment {
    @Override
    public View onCreateView(LayoutInflater inflater,
    ViewGroup container, Bundle savedInstanceState) {
        // Inflate the layout for this fragment
        return inflater.inflate(
            R.layout.fragment2, container, false);
    }
}
```

8. Select the `Fragment` project in Eclipse and press F11 to debug the application on the Android emulator. Figure 2-10 shows the two fragments contained within the activity.

FIGURE 2-10

How It Works

A fragment behaves very much like an activity — it has a Java class and it loads its UI from an XML file. The XML file contains all the usual UI elements that you expect from an activity — `TextView`, `EditText`, `Button`, and so on. The Java class for a fragment needs to extend the `Fragment` base class:

 NOTE *Besides the* `Fragment` *base class, a fragment can also extend a few other subclasses of the* `Fragment` *class, such as* `DialogFragment`, `ListFragment`, *and* `PreferenceFragment`. *Chapter 3 will discuss these types of fragments in more detail.*

```
public class Fragment1 extends Fragment {
    //...
}
```

To draw the UI for a fragment, you override the `onCreateView()` method. This method needs to return a `View` object, like this:

```
public class Fragment1 extends Fragment {
    @Override
    public View onCreateView(LayoutInflater inflater,
    ViewGroup container, Bundle savedInstanceState) {
        // Inflate the layout for this fragment
        return inflater.inflate(
            R.layout.fragment1, container, false);
    }
}
```

Here, you use a `LayoutInflater` object to inflate the UI from the specified XML file (`R.layout .fragment1` in this case). The `container` argument refers to the parent `ViewGroup`, which is the activity in which you are trying to embed the fragment. The `savedInstanceState` argument enables you to restore the fragment to its previously saved state.

To add a fragment to an activity, you use the `<fragment>` element:

```
<LinearLayout xmlns:android="http://schemas.android.com/apk/res/android"
    android:orientation="horizontal"
    android:layout_width="match_parent"
    android:layout_height="match_parent">
    <fragment
        android:name="net.learn2develop.Fragments.Fragment1"
        android:id="@+id/fragment1"
        android:layout_weight="1"
        android:layout_width="0px"
        android:layout_height="match_parent" />
    <fragment
        android:name="net.learn2develop. Fragments.Fragment2"
        android:id="@+id/fragment2"
```

```
        android:layout_weight="1"
        android:layout_width="0px"
        android:layout_height="match_parent" />
</LinearLayout>
```

Note that each fragment needs a unique identifier. You can assign one via the android:id or android:tag attribute.

Adding Fragments Dynamically

While fragments enable you to compartmentalize your UI into various configurable parts, the real power of fragments is realized when you add them dynamically to activities during runtime. In the previous section, you saw how you added fragments to an activity by modifying the XML file during design time. In reality, it is much more useful if you create fragments and add them to activities during runtime. This allows you to create a customizable user interface for your application. For example, if the application is running on a smartphone, you might fill an activity with a single fragment; if the application is running on a tablet, you might then fill the activity with two or more fragments, since the tablet has a much bigger screen estate compared to a smartphone.

TRY IT OUT Adding Fragments during Runtime

1. Using the same project created in the previous section, modify the main.xml file by commenting out the two <fragment> elements:

```xml
<?xml version="1.0" encoding="utf-8"?>
<LinearLayout
    xmlns:android="http://schemas.android.com/apk/res/android"
    android:orientation="horizontal"
    android:layout_width="match_parent"
    android:layout_height="match_parent">
    <!--
    <fragment
        android:name="net.learn2develop.Fragments.Fragment1"
        android:id="@+id/fragment1"
        android:layout_weight="1"
        android:layout_width="0px"
        android:layout_height="match_parent" />

    <fragment
        android:name="net.learn2develop.Fragments.Fragment2"
        android:id="@+id/fragment2"
        android:layout_weight="1"
        android:layout_width="0px"
        android:layout_height="match_parent" />
    -->
</LinearLayout>
```

2. Add the following code in bold to the `MainActivity.java` file:

```
package net.learn2develop.Fragments;

import net.learn2develop.Fragments.R;
import android.app.Activity;
import android.os.Bundle;

import android.app.FragmentManager;
import android.app.FragmentTransaction;
import android.view.Display;
import android.view.WindowManager;

public class MainActivity extends Activity {
    /** Called when the activity is first created. */
    @Override
    public void onCreate(Bundle savedInstanceState) {
        super.onCreate(savedInstanceState);
        setContentView(R.layout.main);

        FragmentManager fragmentManager = getFragmentManager();
        FragmentTransaction fragmentTransaction =
            fragmentManager.beginTransaction();

        //---get the current display info---
        WindowManager wm = getWindowManager();
        Display d = wm.getDefaultDisplay();
        if (d.getWidth() > d.getHeight())
        {
            //---landscape mode---
            Fragment1 fragment1 = new Fragment1();
            // android.R.id.content refers to the content
            // view of the activity
            fragmentTransaction.replace(
                android.R.id.content, fragment1);
        }
        else
        {
            //---portrait mode---
            Fragment2 fragment2 = new Fragment2();
            fragmentTransaction.replace(
                android.R.id.content, fragment2);
        }
        fragmentTransaction.commit();
    }
}
```

3. Press F11 to run the application on the Android emulator. Observe that when the emulator is in landscape mode, fragment 1 (green) is displayed (as Figure 2-11). If you press Ctrl+F11 to change the orientation of the emulator to landscape, fragment 2 (yellow) is added instead (see Figure 2-12).

FIGURE 2-11

FIGURE 2-12

How It Works

To add fragments to an activity, you use the `FragmentManager` class by first obtaining an instance of it:

```
FragmentManager fragmentManager = getFragmentManager();
```

You also need to use the `FragmentTransaction` class to perform fragment transactions in your activity (such as add, remove or replace):

```
FragmentTransaction fragmentTransaction =
        fragmentManager.beginTransaction();
```

In this example, you used the `WindowManager` to determine whether the device is currently in portrait or landscape mode. Once that is determined, you add the appropriate fragment to the activity by creating the fragment and then calling the `replace()` method of the `FragmentTransaction` object to add the fragment to the specified view container (in this case, `android.R.id.content` refers to the content view of the activity):

```
//---landscape mode---
Fragment1 fragment1 = new Fragment1();
// android.R.id.content refers to the content
// view of the activity
fragmentTransaction.replace(
        android.R.id.content, fragment1);
```

Using the `replace()` method is essentially the same as calling the `remove()` method followed by the `add()` method of the `FragmentTransaction` object. To ensure that the changes take effect, you need to call the `commit()` method.

Understanding the Life Cycle of a Fragment

Like activities, fragments have their own life cycle. Understanding the life cycle of a fragment enables you to properly save an instance of the fragment when it is destroyed, and restore it to its previous state when it is recreated.

The following Try It Out examines the various states of a fragment.

TRY IT OUT **Working through the Life Cycle of a Fragment**

1. Using the same project created in the previous section, add the following code in bold to the `Fragment1.java` file:

```
package net.learn2develop.Fragments;

import android.app.Activity;
import android.app.Fragment;
import android.os.Bundle;
import android.view.LayoutInflater;
```

```java
import android.view.View;
import android.view.ViewGroup;

import android.util.Log;

public class Fragment1 extends Fragment {

    @Override
    public void onAttach(Activity activity) {
        super.onAttach(activity);
        Log.d("Fragment 1", "onAttach");
    }

    @Override
    public void onCreate(Bundle savedInstanceState) {
        super.onCreate(savedInstanceState);
        Log.d("Fragment 1", "onCreate");
    }

    @Override
    public View onCreateView(LayoutInflater inflater,
    ViewGroup container, Bundle savedInstanceState) {
        Log.d("Fragment 1", "onCreateView");

        // Inflate the layout for this fragment
        return inflater.inflate(R.layout.fragment1,
            container, false);
    }

    @Override
    public void onActivityCreated(Bundle savedInstanceState) {
        super.onActivityCreated(savedInstanceState);
        Log.d("Fragment 1", "onActivityCreated");
    }

    @Override
    public void onStart() {
        super.onStart();
        Log.d("Fragment 1", "onStart");
    }

    @Override
    public void onResume() {
        super.onResume();
        Log.d("Fragment 1", "onResume");
    }

    @Override
    public void onPause() {
        super.onPause();
        Log.d("Fragment 1", "onPause");
    }

    @Override
```

```java
public void onStop() {
    super.onStop();
    Log.d("Fragment 1", "onStop");
}

@Override
public void onDestroyView() {
    super.onDestroyView();
    Log.d("Fragment 1", "onDestroyView");
}

@Override
public void onDestroy() {
    super.onDestroy();
    Log.d("Fragment 1", "onDestroy");
}

@Override
public void onDetach() {
    super.onDetach();
    Log.d("Fragment 1", "onDetach");
}
}
```

2. Switch the Android emulator to landscape mode by pressing Ctrl+F11.

3. Press F11 in Eclipse to debug the application on the Android emulator.

4. When the application is loaded on the emulator, the following is displayed in the LogCat window (Windows ⇨ Show View ⇨ LogCat):

```
04-02 06:55:38.953: DEBUG/Fragment 1(6818): onAttach
04-02 06:55:38.953: DEBUG/Fragment 1(6818): onCreate
04-02 06:55:38.963: DEBUG/Fragment 1(6818): onCreateView
04-02 06:55:38.983: DEBUG/Fragment 1(6818): onActivityCreated
04-02 06:55:38.983: DEBUG/Fragment 1(6818): onStart
04-02 06:55:39.003: DEBUG/Fragment 1(6818): onResume
```

5. Press the Home button on the emulator. The following output will be displayed in the LogCat window:

```
04-02 04:03:45.543: DEBUG/Fragments(2606): onPause
04-02 04:03:47.394: DEBUG/Fragments(2606): onStop
```

6. On the emulator, click the Apps button in the top-right corner of the screen to launch the application again. This time, the following is displayed:

```
04-02 04:04:32.703: DEBUG/Fragments(2606): onStart
04-02 04:04:32.703: DEBUG/Fragments(2606): onResume
```

7. Finally, click the Back button on the emulator. Now you should see the following output:

```
04-02 07:23:07.393: DEBUG/Fragment 1(7481): onPause
04-02 07:23:07.393: DEBUG/Fragment 1(7481): onStop
```

```
04-02 07:23:07.393: DEBUG/Fragment 1(7481): onDestroyView
04-02 07:23:07.403: DEBUG/Fragment 1(7481): onDestroy
04-02 07:23:07.413: DEBUG/Fragment 1(7481): onDetach
```

How It Works

Like activities, fragments in Android also have their own life cycle. As you have seen, when a fragment is being created, it goes through the following states:

➤   onAttach()

➤   onCreate()

➤   onCreateView()

➤   onActivityCreated()

When the fragment becomes visible, it goes through these states:

➤   onStart()

➤   onResume()

When the fragment goes into the background mode, it goes through these states:

➤   onPause()

➤   onStop()

When the fragment is destroyed (when the activity it is currently hosted in is destroyed), it goes through the following states:

➤   onPause()

➤   onStop()

➤   onDestroyView()

➤   onDestroy()

➤   onDetach()

Like activities, you can restore an instance of a fragment using a Bundle object, in the following states:

➤   onCreate()

➤   onCreateView()

➤   onActivityCreated()

 NOTE *You can save a fragment's state in the* onSaveInstanceState() *event.*

Most of the states experienced by a fragment are similar to those of activities. However, a few new states are specific to fragments:

➤ onAttached() — Called when the fragment has been associated with the activity

➤ onCreateView() — Called to create the view for the fragment

➤ onActivityCreated() — Called when the activity's onCreate() method has been returned

➤ onDestroyView() — Called when the fragment's view is being removed

➤ onDetach() — Called when the fragment is detached from the activity

Note one of the main differences between activities and fragments. When an activity goes into the background, the activity is placed in the back stack. This allows the activity to be resumed when the user presses the Back button. In the case of fragments, however, they are not automatically placed in the back stack when they go into the background. Rather, to place a fragment into the back stack, you need to explicitly call the addToBackStack() method during a fragment transaction, like this:

```
if (d.getWidth() > d.getHeight())
{
    //---landscape mode---
    Fragment1 fragment1 = new Fragment1();
    fragmentTransaction.replace(
        R.id.fragmentContainer, fragment1);
}
else
{
    //---portrait mode---
    Fragment2 fragment2 = new Fragment2();
    fragmentTransaction.replace(
        R.id.fragmentContainer, fragment2);
}

//---add to the back stack---
fragmentTransaction.addToBackStack(null);
fragmentTransaction.commit();
```

The preceding code ensures that after the fragment has been added to the activity, the user can click the Back button to remove it.

Interactions between Fragments

Very often, an activity may contain one or more fragments working together to present a coherent UI to the user. In this case, it is very important for fragments to communicate with one another and exchange data. For example, one fragment might contain a list of items (such as postings from a RSS feed) and when the user taps on an item in that fragment, the details of the selected item may be displayed in another fragment.

The following Try It Out shows how one fragment can access the views contained within another fragment.

TRY IT OUT Communication between Fragments

1. Using the same project created in the previous section, add the following statement in bold to the Fragment1.xml file:

```
<?xml version="1.0" encoding="utf-8"?>
<LinearLayout
    xmlns:android="http://schemas.android.com/apk/res/android"
    android:orientation="vertical"
    android:layout_width="fill_parent"
    android:layout_height="fill_parent"
    android:background="#00FF00"
    >
<TextView
    android:id="@+id/lblFragment1"
    android:layout_width="fill_parent"
    android:layout_height="wrap_content"
    android:text="This is fragment #1" />
</LinearLayout>
```

2. Add the following lines in bold to fragment2.xml:

```
<?xml version="1.0" encoding="utf-8"?>
<LinearLayout
    xmlns:android="http://schemas.android.com/apk/res/android"
    android:orientation="vertical"
    android:layout_width="fill_parent"
    android:layout_height="fill_parent"
    android:background="#FFFE00"
    >
<TextView
    android:layout_width="fill_parent"
    id:layout_height="wrap_content"
    id:text="This is fragment #2" />

    ndroid:id="@+id/btnGetText"
    ndroid:layout_width="wrap_content"
    ndroid:layout_height="wrap_content"
    ndroid:text="Get text in Fragment #1" />

    out>
```

Modify the MainActivity.java file by commenting out the code that you have added in the earlier sections. It should look like this after modification:

```
package net.learn2develop.Fragments;

import net.learn2develop.Fragments.R;
import android.app.Activity;
import android.os.Bundle;

public class MainActivity extends Activity {
    /** Called when the activity is first created. */
    @Override
```

```
        public void onCreate(Bundle savedInstanceState) {
            super.onCreate(savedInstanceState);
            setContentView(R.layout.main);
        }
    }
```

4. Add the following statements in bold to the `Fragment2.java` file:

```
package net.learn2develop.Fragments;

import android.app.Fragment;
import android.os.Bundle;
import android.view.LayoutInflater;
import android.view.View;
import android.view.ViewGroup;
import android.widget.Button;
import android.widget.TextView;
import android.widget.Toast;

public class Fragment2 extends Fragment {
    @Override
    public View onCreateView(LayoutInflater inflater,
    ViewGroup container, Bundle savedInstanceState) {
        // Inflate the layout for this fragment
        return inflater.inflate(R.layout.fragment2,
            container, false);
    }

    @Override
    public void onStart() {
        super.onStart();
        //---Button view---
        Button btnGetText = (Button)
            getActivity().findViewById(R.id.btnGetText);
        btnGetText.setOnClickListener(new View.OnClickListener() {
            public void onClick(View v) {
                TextView lbl = (TextView)
                    getActivity().findViewById(R.id.lblFragment1);
                Toast.makeText(getActivity(), lbl.getText(),
                    Toast.LENGTH_SHORT).show();
            }
        });
    }
}
```

5. Put back the two fragments in `main.xml`:

```
<?xml version="1.0" encoding="utf-8"?>
<LinearLayout
    xmlns:android="http://schemas.android.com/apk/res/android"
    android:orientation="horizontal"
    android:layout_width="match_parent"
    android:layout_height="match_parent">
```

```
<fragment
    android:name="net.learn2develop.Fragments.Fragment1"
    android:id="@+id/fragment1"
    android:layout_weight="1"
    android:layout_width="0px"
    android:layout_height="match_parent" />
<fragment
    android:name="net.learn2develop.Fragments.Fragment2"
    android:id="@+id/fragment2"
    android:layout_weight="1"
    android:layout_width="0px"
    android:layout_height="match_parent" />
</LinearLayout>
```

6. Press F11 to debug the application on the Android emulator. In the second fragment on the right, click the button. You should see the Toast class displaying the text "This is fragment #1" (see Figure 2-13).

FIGURE 2-13

How It Works

As fragments are embedded within activities, you can obtain the activity in which a fragment is currently embedded by using the getActivity() method and then using the findViewById() method to locate the view(s) contained within the fragment:

```
TextView lbl = (TextView)
    getActivity().findViewById(R.id.lblFragment1);
Toast.makeText(getActivity(), lbl.getText(),
    Toast.LENGTH_SHORT).show();
```

UTILIZING THE ACTION BAR

Besides fragments, another new feature introduced in Android 3.0 is the Action Bar. In place of the traditional title bar located at the top of the device's screen, the Action Bar displays the application icon together with the activity title. Optionally, on the right side of the Action Bar are *action items*.

Figure 2-14 shows the built-in Email application displaying the application icon, activity title, and some action items in the Action Bar. The next section discusses action items in more details.

FIGURE 2-14

The following Try It Out shows how you can programmatically hide or display the Action Bar.

TRY IT OUT Showing and Hiding the Action Bar

1. Using Eclipse, create a new Android project and name it **MyActionBar**.

2. Press F11 to debug the application on the Android emulator. You should see the application and its Action Bar located at the top of the screen (containing the application icon and the application name "MyActionBar"; see Figure 2-15).

FIGURE 2-15

3. To hide the Action Bar, add the following line in bold to the AndroidManifest.xml file:

```xml
<?xml version="1.0" encoding="utf-8"?>
<manifest xmlns:android="http://schemas.android.com/apk/res/android"
    package="net.learn2develop.MyActionBar"
    android:versionCode="1"
    android:versionName="1.0">
```

```
            <uses-sdk android:minSdkVersion="11" />

    <application android:icon="@drawable/icon"
            android:label="@string/app_name">
            <activity android:name=".MainActivity"
                    android:label="@string/app_name"
                    android:theme=
                        "@android:style/Theme.Holo.NoActionBar">
                <intent-filter>
                    <action android:name="android.intent.action.MAIN" />
                    <category
                        android:name="android.intent.category.LAUNCHER" />
                </intent-filter>
            </activity>
    </application>
    </manifest>
```

4. Press F11 to debug the application on the Android emulator again. This time, the Action Bar is not displayed (see Figure 2-16).

FIGURE 2-16

5. You can also programmatically remove the Action Bar using the `ActionBar` class. To do so, you first need to remove the `android:theme` attribute you added in the previous step (note the strikethrough):

```
        <activity android:name=".MainActivity"
                android:label="@string/app_name"
                android:theme=
                    "@android:style/Theme.Holo.NoActionBar">
```

```
                    <intent-filter>
                        <action android:name="android.intent.action.MAIN" />
                        <category
                            android:name="android.intent.category.LAUNCHER" />
                    </intent-filter>
                </activity>
```

6. Modify the `MainActivity.java` file as follows:

```
package net.learn2develop.MyActionBar;

import android.app.Activity;
import android.os.Bundle;
import android.app.ActionBar;

public class MainActivity extends Activity {
    /** Called when the activity is first created. */
    @Override
    public void onCreate(Bundle savedInstanceState) {
        super.onCreate(savedInstanceState);
        setContentView(R.layout.main);

        ActionBar actionBar = getActionBar();
        actionBar.hide();
        //actionBar.show(); //---show it again---
    }
}
```

7. Press F11 to debug the application on the emulator again. The Action Bar remains hidden.

How It Works

The `android:theme` attribute lets you turn off the display of the Action Bar for your activity. Setting this attribute to "`@android:style/Theme.Holo.NoActionBar`" hides the Action Bar. Alternatively, you can programmatically get a reference to the Action Bar during runtime by using the `getActionBar()` method. Calling the `hide()` method hides the Action Bar, and calling the `show()` method displays it.

Note that if you use the `android:theme` attribute to turn off the Action Bar, calling the `getActionBar()` method returns a `null` during runtime. Hence, it is always better to turn the Action Bar on/off programmatically using the `ActionBar` class.

Adding Action Items to the Action Bar

Besides displaying the application icon and the activity title on the left of the Action Bar, you can also display additional items on the Action Bar. These additional items are called *action items*. Action items are shortcuts to some of the commonly performed operations in your application. For example, you might be building an RSS reader application and hence some of the action items might be "Refresh feed," "Delete feed" and "Add new feed."

The following Try It Out shows how you can add action items to the Action Bar.

TRY IT OUT **Adding Action Items**

1. Using the same project created in the previous section, add the following code in bold to the MainActivity.java file:

```java
package net.learn2develop.MyActionBar;

import android.app.Activity;
import android.os.Bundle;
import android.app.ActionBar;

import android.view.Menu;
import android.view.MenuItem;
import android.widget.Toast;

public class MainActivity extends Activity {
    /** Called when the activity is first created. */
    @Override
    public void onCreate(Bundle savedInstanceState) {
        super.onCreate(savedInstanceState);
        setContentView(R.layout.main);

        ActionBar actionBar = getActionBar();
        //actionBar.hide();
        //actionBar.show(); //---show it again---
    }

    @Override
    public boolean onCreateOptionsMenu(Menu menu) {
        super.onCreateOptionsMenu(menu);
        CreateMenu(menu);
        return true;
    }

    @Override
    public boolean onOptionsItemSelected(MenuItem item)
    {
        return MenuChoice(item);
    }

    private void CreateMenu(Menu menu)
    {
        MenuItem mnu1 = menu.add(0, 0, 0, "Item 1");
        {
            mnu1.setAlphabeticShortcut('a');
            mnu1.setIcon(R.drawable.icon);
        }
        MenuItem mnu2 = menu.add(0, 1, 1, "Item 2");
        {
            mnu2.setAlphabeticShortcut('b');
            mnu2.setIcon(R.drawable.icon);
        }
        MenuItem mnu3 = menu.add(0, 2, 2, "Item 3");
```

```
        {
            mnu3.setAlphabeticShortcut('c');
            mnu3.setIcon(R.drawable.icon);
        }
        MenuItem mnu4 = menu.add(0, 3, 3, "Item 4");
        {
            mnu4.setAlphabeticShortcut('d');
        }
        menu.add(0, 3, 3, "Item 5");
        menu.add(0, 3, 3, "Item 6");
        menu.add(0, 3, 3, "Item 7");
    }

    private boolean MenuChoice(MenuItem item)
    {
        switch (item.getItemId()) {
        case 0:
            Toast.makeText(this, "You clicked on Item 1",
                Toast.LENGTH_LONG).show();
            return true;
        case 1:
            Toast.makeText(this, "You clicked on Item 2",
                Toast.LENGTH_LONG).show();
            return true;
        case 2:
            Toast.makeText(this, "You clicked on Item 3",
                Toast.LENGTH_LONG).show();
            return true;
        case 3:
            Toast.makeText(this, "You clicked on Item 4",
                Toast.LENGTH_LONG).show();
            return true;
        case 4:
            Toast.makeText(this, "You clicked on Item 5",
                Toast.LENGTH_LONG).show();
            return true;
        case 5:
            Toast.makeText(this, "You clicked on Item 6",
                Toast.LENGTH_LONG).show();
            return true;
        case 6:
            Toast.makeText(this, "You clicked on Item 7",
                Toast.LENGTH_LONG).show();
            return true;
        }
        return false;
    }

}
```

2. Press F11 to debug the application on the Android emulator. Observe the icon on the right side of the Action Bar (see Figure 2-17). This is known as the *overflow* action item.

FIGURE 2-17

3. Clicking the overflow action item reveals a list of menus items (see Figure 2-18). Clicking each menu item will cause the Toast class to display the name of the menu item selected.

FIGURE 2-18

How It Works

The Action Bar populates its action items by calling the onCreateOptionsMenu() method of an activity:

```
@Override
public boolean onCreateOptionsMenu(Menu menu) {
    super.onCreateOptionsMenu(menu);
    CreateMenu(menu);
    return true;
}
```

In the preceding example, you call the CreateMenu() method to display a list of menu items:

```
private void CreateMenu(Menu menu)
{
    MenuItem mnu1 = menu.add(0, 0, 0, "Item 1");
    {
        mnu1.setAlphabeticShortcut('a');
        mnu1.setIcon(R.drawable.icon);
    }
    MenuItem mnu2 = menu.add(0, 1, 1, "Item 2");
    {
        mnu2.setAlphabeticShortcut('b');
        mnu2.setIcon(R.drawable.icon);
    }
    MenuItem mnu3 = menu.add(0, 2, 2, "Item 3");
    {
        mnu3.setAlphabeticShortcut('c');
        mnu3.setIcon(R.drawable.icon);
    }
    MenuItem mnu4 = menu.add(0, 3, 3, "Item 4");
    {
        mnu4.setAlphabeticShortcut('d');
    }
    menu.add(0, 3, 3, "Item 5");
    menu.add(0, 3, 3, "Item 6");
    menu.add(0, 3, 3, "Item 7");
}
```

When a menu item is selected by the user, the onOptionsItemSelected() method is called:

```
@Override
public boolean onOptionsItemSelected(MenuItem item)
{
    return MenuChoice(item);
}
```

Here, you call the self-defined MenuChoice() method to check which menu item was clicked and then print out a message:

```
private boolean MenuChoice(MenuItem item)
{
    switch (item.getItemId()) {
    case 0:
        Toast.makeText(this, "You clicked on Item 1",
            Toast.LENGTH_LONG).show();
        return true;
    case 1:
        Toast.makeText(this, "You clicked on Item 2",
            Toast.LENGTH_LONG).show();
        return true;
    case 2:
        Toast.makeText(this, "You clicked on Item 3",
            Toast.LENGTH_LONG).show();
        return true;
    case 3:
        Toast.makeText(this, "You clicked on Item 4",
            Toast.LENGTH_LONG).show();
        return true;
    case 4:
        Toast.makeText(this, "You clicked on Item 5",
            Toast.LENGTH_LONG).show();
        return true;
    case 5:
        Toast.makeText(this, "You clicked on Item 6",
            Toast.LENGTH_LONG).show();
        return true;
    case 6:
        Toast.makeText(this, "You clicked on Item 7",
            Toast.LENGTH_LONG).show();
        return true;
    }
    return false;
}
```

By default, all the menu items are grouped and displayed under the overflow action button, which is on the far right side of the Action Bar.

Customizing the Action Items and Application Icon

From the previous Try It Out, note that even though the first three menu items have their icons set, they are not displayed in the overflow action item. To make the icon for each menu item appear, you have to make the menu item appear as an action item, not within the overflow action item.

Suppose you have an image named save.png located in each of the drawable folders in the res folder. You can modify the previous Try It Out as follows:

```
MenuItem mnu1 = menu.add(0, 0, 0, "Item 1");
{
    mnu1.setAlphabeticShortcut('a');
```

```
        mnu1.setIcon(R.drawable.save);
        mnu1.setShowAsAction(MenuItem.SHOW_AS_ACTION_IF_ROOM);
    }
```

Doing so causes the menu item to appear as an action item (see Figure 2-19). The SHOW_AS_ACTION_
IF_ROOM constant tells the device to display the menu item as an action item if there is room for
it. This is because you may run out of room to display the action item if too many menu items are
vying for space in the Action Bar. In general, you should restrict the number of action items in the
Action Bar to prevent overcrowding — three should be the maximum.

FIGURE 2-19

If you want to display the text for the action item together with the icon, you could use the "|"
operator together with the MenuItem.SHOW_AS_ACTION_WITH_TEXT constant:

```
        MenuItem mnu1 = menu.add(0, 0, 0, "Item 1");
        {
            mnu1.setAlphabeticShortcut('a');
            mnu1.setIcon(R.drawable.save);
            mnu1.setShowAsAction(MenuItem.SHOW_AS_ACTION_IF_ROOM |
                            MenuItem.SHOW_AS_ACTION_WITH_TEXT);
        }
```

This causes the icon to be displayed together with the text of the menu item (see Figure 2-20).

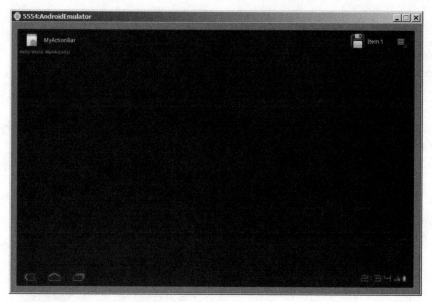

FIGURE 2-20

Besides clicking the action items, users can also click the application icon on the Action Bar. When the application icon is clicked, the onOptionsItemSelected() method is called. To identify the application icon being called, you check the item id against the android.R.id.home constant:

```
private boolean MenuChoice(MenuItem item)
{
    switch (item.getItemId()) {
    case   android.R.id.home:
        Toast.makeText(this,
            "You clicked on the Application icon",
            Toast.LENGTH_LONG).show();
        return true;
    case 0:
        Toast.makeText(this, "You clicked on Item 1",
            Toast.LENGTH_LONG).show();
        return true;
    //...
    //...
    return false;
}
```

The application icon is often used by applications to enable them to return to the main activity of the application. For example, your application may have several activities, and you can use the application icon as a shortcut for users to return directly to the main activity of your application. To do this, it is always good practice to create an Intent object and set it using the Intent.FLAG_ACTIVITY_CLEAR_TOP flag:

```
case android.R.id.home:
    Toast.makeText(this,
```

```
            "You clicked on the Application icon",
            Toast.LENGTH_LONG).show();
            Intent i = new Intent(this, MainActivity.class);
            i.addFlags(Intent.FLAG_ACTIVITY_CLEAR_TOP);
            startActivity(i);
            return true;
```

The `Intent.FLAG_ACTIVITY_CLEAR_TOP` flag ensures that the series of activities in the back stack is cleared when the user clicks the application icon on the Action Bar. This way, if the user clicks the Back button, the other activities in the application do not appear again.

SUMMARY

In this chapter, you have learned about the three most important components of an Android 3.0 application: activities, fragments, and the Action Bar.

An activity defines the UI of your application, whereas a fragment breaks down an activity into smaller manageable chunks. Depending on the device that the application is currently running on, your application can show or hide different fragments, enabling your application to display the best UI for the current device.

Along with Android 3.0, activities now have the Action Bar, which enables them to display commonly used items, such as options menu items. In the following chapters, you will have a chance to see all these new features in action.

EXERCISES

1. Name the two ways to add fragments to an activity.

2. Name one key difference between a fragment and an activity.

3. How do you add action items to an Action Bar?

Answers to the Exercises can be found in Appendix C.

▶ WHAT YOU LEARNED IN THIS CHAPTER

TOPIC	KEY CONCEPTS
Activity	Contains the UI of your Android application.
Life cycle of an activity	An activity is destroyed when the user presses the Back button. Otherwise, it goes into the background when it loses visibility. To preserve the state of an activity, handle its `onPause()` event and restore it in the `onStart()` or `onResume()` events.
Fragment	A fragment is a mini-activity, with its own life cycle. Fragments are embedded in activities.
Manipulating fragments programmatically	You need to use the `FragmentManager` and `FragmentTransaction` classes when adding, removing, or replacing fragments during runtime.
Life cycle of a fragment	Similar to that of an activity — you save the state of a fragment in the `onPause()` event, and restore its state in one of the following events: `onCreate()`, `onCreateView()`, or `onActivityCreated()`.
Action Bar	Replaces the traditional title bar for older versions of Android.
Action items	Action items are displayed on the right of the Action Bar. They are created just like options menus.
Application icon	Usually used for going back to the "home" activity of an application. It is advisable to use the `Intent` object with the `Intent.FLAG_ACTIVITY_CLEAR_TOP` flag.

3

Android User Interface

WHAT YOU WILL LEARN IN THIS CHAPTER

➤ What the various ViewGroups are that you can use to lay out your views

➤ How to use the basic views in Android to design your user interface

➤ How to use the specialized fragments available in Android 3.0

In Chapter 2, you learned about the `Activity` and `Fragment` classes and their life cycles. You learned that an activity (as well as a fragment) is a means by which users interact with the application. However, an activity or fragment by itself does not have a presence on the screen. Instead, it has to draw the screen using *Views* and *ViewGroups*. In this chapter, you learn the details about creating user interfaces in Android, and how users interact with them.

VIEWS AND VIEWGROUPS

An activity or fragment contains Views and ViewGroups. A view is a widget that has an appearance on screen. Examples of views are buttons, labels, and text boxes. A view derives from the base class `android.view.View`.

One or more views can be grouped together into a ViewGroup. A ViewGroup (which is itself a special type of view) provides the layout in which you can order the appearance and sequence of views. Examples of ViewGroups include `LinearLayout` and `FrameLayout`. A ViewGroup derives from the base class `android.view.ViewGroup`.

Android supports the following ViewGroups:

➤ `LinearLayout`

➤ `AbsoluteLayout`

➤ TableLayout

➤ RelativeLayout

➤ FrameLayout

➤ ScrollView

The following sections describe each of these ViewGroups in more detail. Note that in practice it is common to combine different types of layouts to create the UI you want.

LinearLayout

The LinearLayout arranges views in a single column or a single row. Child views can be arranged either vertically or horizontally. To see how LinearLayout works, consider the following elements typically contained in the main.xml file:

```xml
<?xml version="1.0" encoding="utf-8"?>
<LinearLayout
    xmlns:android="http://schemas.android.com/apk/res/android"
    android:orientation="vertical"
    android:layout_width="fill_parent"
    android:layout_height="fill_parent"
    >
<TextView
    android:layout_width="fill_parent"
    android:layout_height="wrap_content"
    android:text="@string/hello"
    />
</LinearLayout>
```

In the main.xml file, observe that the root element is <LinearLayout> and it has a <TextView> element contained within it. The <LinearLayout> element controls the order in which the views contained within it appear.

Each View and ViewGroup has a set of common attributes, some of which are described in Table 3-1.

TABLE 3-1: Common Attributes Used in Views and ViewGroups

ATTRIBUTE	DESCRIPTION
layout_width	Specifies the width of the View or ViewGroup
layout_height	Specifies the height of the View or ViewGroup
layout_marginTop	Specifies extra space on the top side of the View or ViewGroup
layout_marginBottom	Specifies extra space on the bottom side of the View or ViewGroup
layout_marginLeft	Specifies extra space on the left side of the View or ViewGroup
layout_marginRight	Specifies extra space on the right side of the View or ViewGroup

ATTRIBUTE	DESCRIPTION
layout_gravity	Specifies how child Views are positioned
layout_weight	Specifies how much of the extra space in the layout should be allocated to the View
layout_x	Specifies the x-coordinate of the View or ViewGroup
layout_y	Specifies the y-coordinate of the View or ViewGroup

 NOTE *Some of these attributes are applicable only when a View is in a specific ViewGroup. For example, the* layout_weight *and* layout_gravity *attributes are applicable only when a View is in either a* LinearLayout *or a* TableLayout.

For example, the width of the <TextView> element fills the entire width of its parent (which is the screen in this case) using the fill_parent constant. Its height is indicated by the wrap_content constant, which means that its height is the height of its content (in this case, the text contained within it). If you don't want to have the <TextView> view occupy the entire row, you can set its layout_width attribute to wrap_content, like this:

```
<TextView
    android:layout_width="wrap_content"
    android:layout_height="wrap_content"
    android:text="@string/hello"
/>
```

This will set the width of the view to be equal to the width of the text contained within it.

UNITS OF MEASUREMENT

When specifying the size of an element on an Android UI, you should be aware of the following units of measurement:

dp — Density-independent pixel. 160dp is equivalent to one inch of physical screen size. This is the recommended unit of measurement when specifying the dimension of views in your layout. You can specify either "dp" or "dip" when referring to a density-independent pixel.

sp — Scale-independent pixel. This is similar to dp and is recommended for specifying font sizes.

pt — Point. A point is defined to be $1/72$ of an inch, based on the physical screen size.

px — Pixel. Corresponds to actual pixels on the screen. Using this unit is not recommended, as your UI may not render correctly on devices with different screen sizes.

Consider the following layout:

```xml
<?xml version="1.0" encoding="utf-8"?>
<LinearLayout xmlns:android="http://schemas.android.com/apk/res/android"
    android:orientation="vertical"
    android:layout_width="fill_parent"
    android:layout_height="fill_parent"
    >
<TextView
    android:layout_width="105dp"
    android:layout_height="wrap_content"
    android:text="@string/hello"
    />
<Button
    android:layout_width="160dp"
    android:layout_height="wrap_content"
    android:text="Button"
    />
</LinearLayout>
```

Here, you set the width of both the `TextView` and `Button` views to an absolute value. In this case, the width for the `TextView` is set to 105 density-independent pixels wide, and the `Button` to 160 density-independent pixels wide. Figure 3-1 shows how the views look when viewed on an emulator with a resolution of 320×480.

Figure 3-2 shows how the views look when viewed on a high-resolution (480×800) emulator.

FIGURE 3-1

FIGURE 3-2

As you can see, in both emulators the width of both views is the same with respect to the width of the emulator. This demonstrates the usefulness of using the dp unit, which ensures that even if the resolution of the target device is different, the size of the view relative to the device remains unchanged.

The preceding example also specifies that the orientation of the layout is vertical:

```
<LinearLayout
    xmlns:android="http://schemas.android.com/apk/res/android"
    android:orientation="vertical"
    android:layout_width="fill_parent"
    android:layout_height="fill_parent"
    >
```

The default orientation layout is horizontal, so if you omit the android:orientation attribute, the views appear as shown in Figure 3-3.

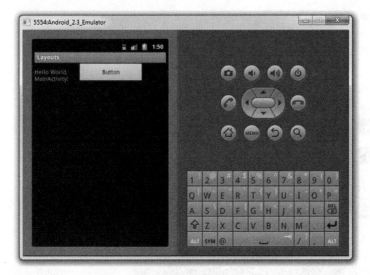

FIGURE 3-3

In LinearLayout, you can apply the layout_weight and layout_gravity attributes to views contained within it, as the following modifications to main.xml show:

```
<?xml version="1.0" encoding="utf-8"?>
<LinearLayout xmlns:android="http://schemas.android.com/apk/res/android"
    android:orientation="vertical"
    android:layout_width="fill_parent"
    android:layout_height="fill_parent"
    >
<TextView
    android:layout_width="105dp"
    android:layout_height="wrap_content"
    android:text="@string/hello"
    />
```

```
<Button
    android:layout_width="160dp"
    android:layout_height="wrap_content"
    android:text="Button"
    android:layout_gravity="right"
    android:layout_weight="0.2"
    />
<EditText
    android:layout_width="fill_parent"
    android:layout_height="wrap_content"
    android:textSize="18sp"
    android:layout_weight="0.8"
    />
</LinearLayout>
```

Figure 3-4 shows that the button is aligned to the right of its parent (which is the LinearLayout) using the layout_gravity attribute. At the same time, you use the layout_weight attribute to specify the ratio in which the Button and EditText views occupy the remaining space on the screen. The total value for the layout_weight attribute must be equal to 1.

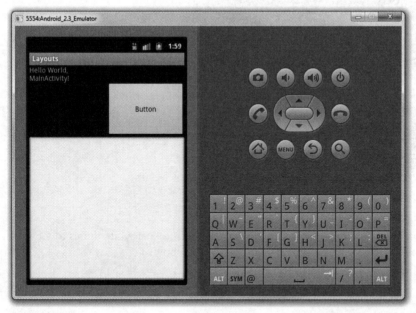

FIGURE 3-4

AbsoluteLayout

The AbsoluteLayout enables you to specify the exact location of its child views. Consider the following UI defined in main.xml:

```
<?xml version="1.0" encoding="utf-8"?>
<AbsoluteLayout
```

```
        android:layout_width="fill_parent"
        android:layout_height="fill_parent"
        xmlns:android="http://schemas.android.com/apk/res/android"
        >
    <Button
        android:layout_width="188dp"
        android:layout_height="wrap_content"
        android:text="Button"
        android:layout_x="126px"
        android:layout_y="361px"
        />
    <Button
        android:layout_width="113dp"
        android:layout_height="wrap_content"
        android:text="Button"
        android:layout_x="12px"
        android:layout_y="361px"
        />
</AbsoluteLayout>
```

Figure 3-5 shows the two `Button` views located at their specified positions using the `android_layout_x` and `android_layout_y` attributes.

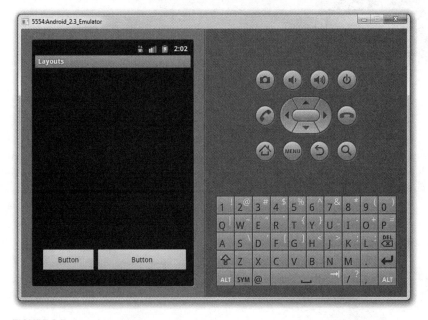

FIGURE 3-5

However, there is a problem with the `AbsoluteLayout` when the activity is viewed on a high-resolution screen (see Figure 3-6). For this reason, the `AbsoluteLayout` has been deprecated since Android 1.5 (although it is still supported in the current version). You should avoid using the `AbsoluteLayout` in your UI, as it is not guaranteed to be supported in future versions of Android. Instead, use the other layouts described in this chapter.

FIGURE 3-6

TableLayout

The `TableLayout` groups views into rows and columns. You use the `<TableRow>` element to designate a row in the table. Each row can contain one or more views. Each view you place within a row forms a cell. The width of each column is determined by the largest width of each cell in that column.

Consider the content of `main.xml` shown here:

```xml
<?xml version="1.0" encoding="utf-8"?>
<TableLayout
    xmlns:android="http://schemas.android.com/apk/res/android"
    android:layout_height="fill_parent"
    android:layout_width="fill_parent"
    >
<TableRow>
    <TextView
        android:text="User Name:"
        android:width ="120px"
        />
    <EditText
        android:id="@+id/txtUserName"
```

```
                    android:width="200px" />
        </TableRow>
        <TableRow>
            <TextView
                android:text="Password:"
                />
            <EditText
                android:id="@+id/txtPassword"
                android:password="true"
                />
        </TableRow>
        <TableRow>
            <TextView />
            <CheckBox android:id="@+id/chkRememberPassword"
                android:layout_width="fill_parent"
                android:layout_height="wrap_content"
                android:text="Remember Password"
                />
        </TableRow>
        <TableRow>
            <Button
                android:id="@+id/buttonSignIn"
                android:text="Log In" />
        </TableRow>
    </TableLayout>
```

Figure 3-7 shows what the preceding looks like when rendered on the Android emulator.

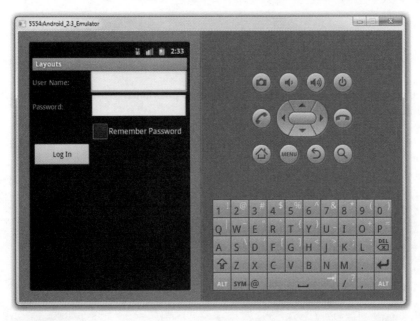

FIGURE 3-7

Note that in the preceding example, there are two columns and four rows in the `TableLayout`. The cell directly under the Password `TextView` is populated with a `<TextView/>` empty element. If you don't do this, the Remember Password checkbox appears under the Password `TextView`, as shown in Figure 3-8.

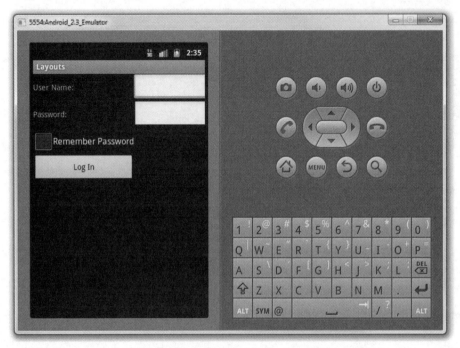

FIGURE 3-8

RelativeLayout

The `RelativeLayout` enables you to specify how child views are positioned relative to each other. Consider the following `main.xml` file:

```
<?xml version="1.0" encoding="utf-8"?>
<RelativeLayout
    android:id="@+id/RLayout"
    android:layout_width="fill_parent"
    android:layout_height="fill_parent"
    xmlns:android="http://schemas.android.com/apk/res/android"
    >
    <TextView
        android:id="@+id/lblComments"
        android:layout_width="wrap_content"
```

```
                android:layout_height="wrap_content"
                android:text="Comments"
                android:layout_alignParentTop="true"
                android:layout_alignParentLeft="true"
                />
        <EditText
                android:id="@+id/txtComments"
                android:layout_width="fill_parent"
                android:layout_height="170px"
                android:textSize="18sp"
                android:layout_alignLeft="@+id/lblComments"
                android:layout_below="@+id/lblComments"
                android:layout_centerHorizontal="true"
                />
        <Button
                android:id="@+id/btnSave"
                android:layout_width="125px"
                android:layout_height="wrap_content"
                android:text="Save"
                android:layout_below="@+id/txtComments"
                android:layout_alignRight="@+id/txtComments"
                />
        <Button
                android:id="@+id/btnCancel"
                android:layout_width="124px"
                android:layout_height="wrap_content"
                android:text="Cancel"
                android:layout_below="@+id/txtComments"
                android:layout_alignLeft="@+id/txtComments"
                />
    </RelativeLayout>
```

Notice that each view embedded within the RelativeLayout has attributes that enable it to align with another view. These attributes are as follows:

- ➤ layout_alignParentTop
- ➤ layout_alignParentLeft
- ➤ layout_alignLeft
- ➤ layout_alignRight
- ➤ layout_below
- ➤ layout_centerHorizontal

The value for each of these attributes is the ID for the view that you are referencing. The preceding XML UI creates the screen shown in Figure 3-9.

FIGURE 3-9

FrameLayout

The `FrameLayout` is an on-screen placeholder that you can use to display a single view. Views that you add to a `FrameLayout` are always anchored to the top left of the layout. Consider the following content in `main.xml`:

```xml
<?xml version="1.0" encoding="utf-8"?>
<RelativeLayout
    android:id="@+id/RLayout"
    android:layout_width="fill_parent"
    android:layout_height="fill_parent"
    xmlns:android="http://schemas.android.com/apk/res/android"
    >
    <TextView
        android:id="@+id/lblComments"
        android:layout_width="wrap_content"
        android:layout_height="wrap_content"
        android:text="This is my lovely dog, Ookii"
        android:layout_alignParentTop="true"
        android:layout_alignParentLeft="true"
        />
    <FrameLayout
        android:layout_width="wrap_content"
        android:layout_height="wrap_content"
        android:layout_alignLeft="@+id/lblComments"
        android:layout_below="@+id/lblComments"
        android:layout_centerHorizontal="true"
        >
```

```
            <ImageView
                android:src = "@drawable/ookii"
                android:layout_width="wrap_content"
                android:layout_height="wrap_content"
                />
        </FrameLayout>
    </RelativeLayout>
```

Here, you have a `FrameLayout` within a `RelativeLayout`. Within the `FrameLayout`, you embed an `ImageView`. The UI is shown in Figure 3-10.

 NOTE *This example assumes that the* `res/drawable-mdpi` *folder has an image named* `ookii.png`.

FIGURE 3-10

If you add another view (such as a `Button` view) within the `FrameLayout`, the view will overlap the previous view (see Figure 3-11):

```
    <?xml version="1.0" encoding="utf-8"?>
    <RelativeLayout
        android:id="@+id/RLayout"
        android:layout_width="fill_parent"
        android:layout_height="fill_parent"
        xmlns:android="http://schemas.android.com/apk/res/android"
        >
        <TextView
```

```
            android:id="@+id/lblComments"
            android:layout_width="wrap_content"
            android:layout_height="wrap_content"
            android:text="This is my lovely dog, Ookii"
            android:layout_alignParentTop="true"
            android:layout_alignParentLeft="true"
            />
    <FrameLayout
            android:layout_width="wrap_content"
            android:layout_height="wrap_content"
            android:layout_alignLeft="@+id/lblComments"
            android:layout_below="@+id/lblComments"
            android:layout_centerHorizontal="true"
            >
        <ImageView
            android:src = "@drawable/ookii"
            android:layout_width="wrap_content"
            android:layout_height="wrap_content"
            />
        <Button
            android:layout_width="124dp"
            android:layout_height="wrap_content"
            android:text="Print Picture"
            />
    </FrameLayout>
</RelativeLayout>
```

FIGURE 3-11

NOTE *You can add multiple views to a* `FrameLayout`, *but each will be stacked on top of the previous one. This is useful in cases where you want to animate series of images, with only one visible at a time.*

ScrollView

A `ScrollView` is a special type of `FrameLayout` in that it enables users to scroll through a list of views that occupy more space than the physical display. The `ScrollView` can contain only one child View or ViewGroup, which normally is a `LinearLayout`.

The following `main.xml` content shows a `ScrollView` containing a `LinearLayout`, which in turn contains some `Button` and `EditText` views:

```xml
<?xml version="1.0" encoding="utf-8"?>
<ScrollView
    android:layout_width="fill_parent"
    android:layout_height="fill_parent"
    xmlns:android="http://schemas.android.com/apk/res/android"
    >
<LinearLayout
    android:layout_width="fill_parent"
    android:layout_height="wrap_content"
    android:orientation="vertical"
    >
    <Button
        android:id="@+id/button1"
        android:layout_width="fill_parent"
        android:layout_height="wrap_content"
        android:text="Button 1"
        />
    <Button
        android:id="@+id/button2"
        android:layout_width="fill_parent"
        android:layout_height="wrap_content"
        android:text="Button 2"
        />
    <Button
        android:id="@+id/button3"
        android:layout_width="fill_parent"
        android:layout_height="wrap_content"
        android:text="Button 3"
        />
    <EditText
```

```
            android:id="@+id/txt"
            android:layout_width="fill_parent"
            android:layout_height="300px"
            />
        <Button
            android:id="@+id/button4"
            android:layout_width="fill_parent"
            android:layout_height="wrap_content"
            android:text="Button 4"
            />
        <Button
            android:id="@+id/button5"
            android:layout_width="fill_parent"
            android:layout_height="wrap_content"
            android:text="Button 5"
            />
    </LinearLayout>
  </ScrollView>
```

Figure 3-12 shows the `ScrollView` enabling users to drag the screen upward to reveal the views located at the bottom of the screen.

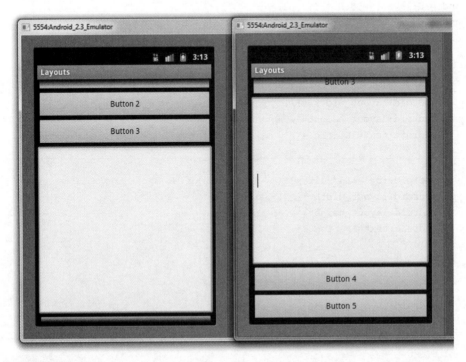

FIGURE 3-12

BASIC VIEWS

In the previous section, you learned about the various layouts that you can use to position your views in an activity. You also learned about the techniques you can use to adapt to different screen resolutions and sizes. In this section, you will take a look at the various views that you can use to design the user interface for your applications.

In particular, you will explore some of the basic views that you can use to design the UI of your Android applications:

➤ TextView

➤ EditText

➤ Button

➤ ImageButton

➤ CheckBox

➤ ToggleButton

➤ RadioButton

➤ RadioGroup

These basic views enable you to display text information, as well as perform some basic selection. The following sections explore all these views in more detail.

TextView View

When you create a new Android project, Eclipse always creates the main.xml file (located in the res/layout folder), which contains a `<TextView>` element:

```xml
<?xml version="1.0" encoding="utf-8"?>
<LinearLayout xmlns:android="http://schemas.android.com/apk/res/android"
    android:orientation="vertical"
    android:layout_width="fill_parent"
    android:layout_height="fill_parent"
    >
<TextView
    android:layout_width="fill_parent"
    android:layout_height="wrap_content"
    android:text="@string/hello"
    />
</LinearLayout>
```

The TextView view is used to display text to the user. This is the most basic view and one that you will frequently use when you develop Android applications. If you need to allow users to edit the text displayed, you should use the subclass of TextView, EditText, which is discussed in the next section.

> **NOTE** *In some other platforms, the* `TextView` *is commonly (though not officially) known as the label view. Its sole purpose is to display text on the screen.*

Common Views

Besides the `TextView` view, which you will likely use the most often, there are some other basic controls that you will find yourself frequently using: `Button`, `ImageButton`, `EditText`, `CheckBox`, `ToggleButton`, `RadioButton`, and `RadioGroup`:

➤ `Button` — Represents a push-button widget.

➤ `ImageButton` — Similar to the Button view, except that it also displays an image.

➤ `EditText` — A subclass of the `TextView` view, except that it allows users to edit its text content.

➤ `CheckBox` — A special type of button that has two states: checked or unchecked.

➤ `RadioGroup` and `RadioButton` — The `RadioButton` has two states: either checked or unchecked. Once a RadioButton is checked, it cannot be unchecked without selecting another RadioButton. A `RadioGroup` is used to group together one or more `RadioButton` views, thereby allowing only one `RadioButton` to be checked within the `RadioGroup`.

➤ `ToggleButton` — Displays checked/unchecked states using a light indicator (green for checked).

The following Try It Out provides details about how these views work.

TRY IT OUT **Using the Basic Views**

codefile BasicViews.zip available for download at Wrox.com

1. Using Eclipse, create an Android project and name it `BasicViews`.

> **NOTE** *For subsequent projects that you will create in this book, the various fields for the project will adopt the following values:*
> **Application Name** — *<project name>*
> **Package name** — *net.learn2develop.<project name>*
> **Create Activity** — *MainActivity*
> **Min SDK Version** — *<version number>*

2. Modify the `main.xml` file located in the `res/layout` folder by adding the following elements shown in bold:

```
<?xml version="1.0" encoding="utf-8"?>
<LinearLayout xmlns:android="http://schemas.android.com/apk/res/
```

```
android"
    android:orientation="vertical"
    android:layout_width="fill_parent"
    android:layout_height="fill_parent">

    <Button android:id="@+id/btnSave"
        android:layout_width="fill_parent"
        android:layout_height="wrap_content"
        android:text="Save" />

    <Button android:id="@+id/btnOpen"
        android:layout_width="wrap_content"
        android:layout_height="wrap_content"
        android:text="Open" />

    <ImageButton android:id="@+id/btnImg1"
        android:layout_width="fill_parent"
        android:layout_height="wrap_content"
        android:src="@drawable/icon" />

    <EditText android:id="@+id/txtName"
        android:layout_width="fill_parent"
        android:layout_height="wrap_content" />

    <CheckBox android:id="@+id/chkAutosave"
        android:layout_width="fill_parent"
        android:layout_height="wrap_content"
        android:text="Autosave" />

    <CheckBox android:id="@+id/star"
        style="?android:attr/starStyle"
        android:layout_width="wrap_content"
        android:layout_height="wrap_content" />

    <RadioGroup android:id="@+id/rdbGp1"
        android:layout_width="fill_parent"
        android:layout_height="wrap_content"
        android:orientation="vertical" >
        <RadioButton android:id="@+id/rdb1"
            android:layout_width="fill_parent"
            android:layout_height="wrap_content"
            android:text="Option 1" />
        <RadioButton android:id="@+id/rdb2"
            android:layout_width="fill_parent"
            android:layout_height="wrap_content"
            android:text="Option 2" />
```

```
    </RadioGroup>

    <ToggleButton android:id="@+id/toggle1"
        android:layout_width="wrap_content"
        android:layout_height="wrap_content" />

</LinearLayout>
```

3. To see the views in action, debug the project in Eclipse by selecting the project name and pressing F11.

Figure 3-13 shows the various views displayed in the Android emulator.

FIGURE 3-13

4. Click on the various views and note how they vary in their look and feel. Figure 3-14 shows the following changes to the view:

- The first CheckBox view (Autosave) is checked.
- The second CheckBox view (star) is checked.
- The second RadioButton (Option 2) is selected.
- The ToggleButton is turned on.

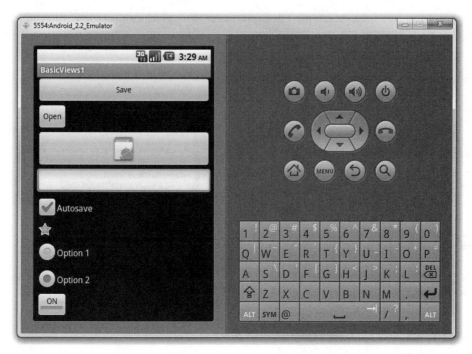

FIGURE 3-14

How It Works

So far, all the views are relatively straightforward — they are listed using the `<LinearLayout>` element, so they are stacked on top of each other when they are displayed in the activity.

For the first `Button`, the `layout_width` attribute is set to `fill_parent` so that its width occupies the entire width of the screen:

```
<Button android:id="@+id/btnSave"
    android:layout_width="fill_parent"
    android:layout_height="wrap_content"
    android:text="Save" />
```

For the second `Button`, the `layout_width` attribute is set to `wrap_content` so that its width will be the width of its content — specifically, the text that it is displaying (i.e.,"Open"):

```
<Button android:id="@+id/btnOpen"
    android:layout_width="wrap_content"
    android:layout_height="wrap_content"
    android:text="Open" />
```

The `ImageButton` displays a button with an image. The image is set through the `src` attribute. In this case, you simply use the image used for the application icon:

```
<ImageButton android:id="@+id/btnImg1"
    android:layout_width="fill_parent"
    android:layout_height="wrap_content"
    android:src="@drawable/icon" />
```

The `EditText` view displays a rectangular region where the user can enter some text. You set the `layout_height` to `wrap_content` so that if the user enters a long string of text, its height will automatically be adjusted to fit the content (see Figure 3-15).

This is a long sentence that will span multiple lines...

Autosave

FIGURE 3-15

```
<EditText android:id="@+id/txtName"
    android:layout_width="fill_parent"
    android:layout_height="wrap_content" />
```

The `CheckBox` displays a checkbox that users can tap to check or uncheck it:

```
<CheckBox android:id="@+id/chkAutosave"
    android:layout_width="fill_parent"
    android:layout_height="wrap_content"
    android:text="Autosave" />
```

If you don't like the default appearance of the `CheckBox`, you can apply a `style` attribute to it to display it as some other image, such as a star:

```
<CheckBox android:id="@+id/star"
    style="?android:attr/starStyle"
    android:layout_width="wrap_content"
    android:layout_height="wrap_content" />
```

The format for the value of the `style` attribute is as follows:

```
?[package:][type:]name.
```

The `RadioGroup` encloses two `RadioButtons`. This is important because radio buttons are usually used to present multiple options to the user for selection. When a `RadioButton` in a `RadioGroup` is selected, all other `RadioButtons` are automatically unselected:

```
<RadioGroup android:id="@+id/rdbGp1"
    android:layout_width="fill_parent"
    android:layout_height="wrap_content"
    android:orientation="vertical" >
<RadioButton android:id="@+id/rdb1"
    android:layout_width="fill_parent"
    android:layout_height="wrap_content"
    android:text="Option 1" />
```

```
    <RadioButton android:id="@+id/rdb2"
        android:layout_width="fill_parent"
        android:layout_height="wrap_content"
        android:text="Option 2" />
</RadioGroup>
```

Notice that the `RadioButtons` are listed vertically, one on top of another. If you want to list them horizontally, you need to change the `orientation` attribute to `horizontal`. You would also need to ensure that the `layout_width` attribute of the `RadioButtons` are set to `wrap_content`:

```
<RadioGroup android:id="@+id/rdbGp1"
    android:layout_width="fill_parent"
    android:layout_height="wrap_content"
    android:orientation="horizontal" >
    <RadioButton android:id="@+id/rdb1"
        android:layout_width="wrap_content"
        android:layout_height="wrap_content"
        android:text="Option 1" />
    <RadioButton android:id="@+id/rdb2"
        android:layout_width="wrap_content"
        android:layout_height="wrap_content"
        android:text="Option 2" />
</RadioGroup>
```

Figure 3-16 shows the `RadioButtons` displayed horizontally.

The `ToggleButton` displays a rectangular button that users can toggle on and off by clicking it:

FIGURE 3-16

```
<ToggleButton android:id="@+id/toggle1"
    android:layout_width="wrap_content"
    android:layout_height="wrap_content" />
```

One thing that has been consistent throughout this example is that each view has the `id` attribute set to a particular value, such as in the case of the `Button`:

```
<Button android:id="@+id/btnSave"
    android:layout_width="fill_parent"
    android:layout_height="wrap_content"
    android:text="Save" />
```

The `id` attribute is an identifier for a view so that it may later be retrieved using the `View` `.findViewById()` or `Activity.findViewById()` methods.

Now that you have seen how the various views look for an activity, the following Try It Out demonstrates how you can programmatically control them.

Handling View Events

1. Using the same project created in the previous Try It Out, modify the `MainActivity.java` file by adding the following statements in bold:

```java
package net.learn2develop.BasicViews1;

import android.app.Activity;
import android.os.Bundle;

import android.view.View;
import android.widget.Button;
import android.widget.CheckBox;
import android.widget.RadioButton;
import android.widget.RadioGroup;
import android.widget.Toast;
import android.widget.ToggleButton;
import android.widget.RadioGroup.OnCheckedChangeListener;

public class MainActivity extends Activity {
    /** Called when the activity is first created. */
    @Override
    public void onCreate(Bundle savedInstanceState) {
        super.onCreate(savedInstanceState);
        setContentView(R.layout.main);

        //---Button view---
        Button btnOpen = (Button) findViewById(R.id.btnOpen);
        btnOpen.setOnClickListener(new View.OnClickListener() {
            public void onClick(View v) {
                DisplayToast("You have clicked the Open button");
            }
        });

        //---Button view---
        Button btnSave = (Button) findViewById(R.id.btnSave);
        btnSave.setOnClickListener(new View.OnClickListener()
        {
            public void onClick(View v) {
                DisplayToast("You have clicked the Save button");
            }
        });

        //---CheckBox---
        CheckBox checkBox = (CheckBox) findViewById(R.id.chkAutosave);
        checkBox.setOnClickListener(new View.OnClickListener()
        {
            public void onClick(View v) {
                if (((CheckBox)v).isChecked())
                    DisplayToast("CheckBox is checked");
                else
                    DisplayToast("CheckBox is unchecked");
            }
```

```
        });

        //---RadioButton---
        RadioGroup radioGroup = (RadioGroup) findViewById(R.id.rdbGp1);
        radioGroup.setOnCheckedChangeListener(new OnCheckedChangeListener()
        {
            public void onCheckedChanged(RadioGroup group, int checkedId) {
                RadioButton rb1 = (RadioButton) findViewById(R.id.rdb1);
                if (rb1.isChecked()) {
                    DisplayToast("Option 1 checked!");
                } else {
                    DisplayToast("Option 2 checked!");
                }
            }
        });

        //---ToggleButton---
        ToggleButton toggleButton =
            (ToggleButton) findViewById(R.id.toggle1);
        toggleButton.setOnClickListener(new View.OnClickListener()
        {
            public void onClick(View v) {
                if (((ToggleButton)v).isChecked())
                    DisplayToast("Toggle button is On");
                else
                    DisplayToast("Toggle button is Off");
            }
        });
    }

    private void DisplayToast(String msg)
    {
        Toast.makeText(getBaseContext(), msg,
                Toast.LENGTH_SHORT).show();
    }
}
```

2. Press F11 to debug the project on the Android emulator.

3. Click on the various views and observe the message displayed in the Toast window.

How It Works

To handle the events fired by each view, you first have to programmatically locate the view that you created during the onCreate() event. You do so using the Activity.findViewById() method, supplying it with the ID of the view:

```
        //---Button view---
        Button btnOpen = (Button) findViewById(R.id.btnOpen);
```

The setOnClickListener() method registers a callback to be invoked later when the view is clicked:

```
        btnOpen.setOnClickListener(new View.OnClickListener() {
            public void onClick(View v) {
```

```
                DisplayToast("You have clicked the Open button");
        }
    });
```

The onClick() method is called when the view is clicked.

To determine the state of the CheckBox, you have to typecast the argument of the onClick() method to a CheckBox and then check its isChecked() method to see if it is checked:

```
    //---CheckBox---
    CheckBox checkBox = (CheckBox) findViewById(R.id.chkAutosave);
    checkBox.setOnClickListener(new View.OnClickListener()
    {
        public void onClick(View v) {
            if (((CheckBox)v).isChecked())
                DisplayToast("CheckBox is checked");
            else
                DisplayToast("CheckBox is unchecked");
        }
    });
```

For RadioButtons, you need to use the setOnCheckedChangeListener() method on the RadioGroup to register a callback to be invoked when the checked RadioButton changes in this group:

```
    //---RadioButton---
    RadioGroup radioGroup = (RadioGroup) findViewById(R.id.rdbGp1);
    radioGroup.setOnCheckedChangeListener(new OnCheckedChangeListener()
    {
        public void onCheckedChanged(RadioGroup group, int checkedId) {
            RadioButton rb1 = (RadioButton) findViewById(R.id.rdb1);
            if (rb1.isChecked()) {
                DisplayToast("Option 1 checked!");
            } else {
                DisplayToast("Option 2 checked!");
            }
        }
    });
```

When a RadioButton is selected, the onCheckedChanged() method is fired. Within it, you locate individual RadioButtons and then call their isChecked() method to determine which RadioButton is selected. Alternatively, the onCheckedChanged() method contains a second argument that contains a unique identifier for the RadioButton selected.

 NOTE *Because this book focuses on developing apps for Tablets, events are not covered in depth. For more information on Events in Android, please see* Beginning Android Application Development *(by Lee, Wrox, 2011).*

FRAGMENTS

In Chapter 2, you learn about the fragment feature that is new in Android 3.0. Using fragments, you can customize the user interface of your Android application by dynamically rearranging fragments to fit within an activity. This enables you to build applications that run on devices with different screen sizes.

As you have learned, fragments are really "mini-activities" that have their own life cycles. To create a fragment, you need a class that extends the `Fragment` base class. In Chapter 2, you learned how to create fragments and add them to your activities. Besides the `Fragment` base class, you can also extend from some other subclasses of the `Fragment` base class to create more specialized fragments. The following sections discuss the three subclasses of `Fragment`: `ListFragment`, `DialogFragment`, and `PreferenceFragment`.

ListFragment

A list fragment is a fragment that contains a `ListView`, displaying a list of items from a data source such as an array or a `Cursor`. A list fragment is very useful, as you may often have one fragment that contains a list of items (such as a list of RSS postings), and another fragment that displays the details of the selected posting. To create a list fragment, you need to extend the `ListFragment` base class.

The following Try It Out shows you how to get started with a list fragment.

TRY IT OUT Creating and Using a List Fragment

codefile ListFragmentExample.zip available for download at Wrox.com

1. Using Eclipse, create an Android 3.0 project and name it `ListFragmentExample`.

2. Add a XML file to the `res/layout` folder and name it `fragment1.xml`.

3. Populate the `fragment1.xml` as follows:

```xml
<?xml version="1.0" encoding="utf-8"?>
<LinearLayout xmlns:android="http://schemas.android.com/apk/res/android"
    android:orientation="vertical"
    android:layout_width="fill_parent"
    android:layout_height="fill_parent">
    <ListView
        android:id="@id/android:list"
        android:layout_width="match_parent"
        android:layout_height="match_parent"
        android:layout_weight="1"
        android:drawSelectorOnTop="false"/>
</LinearLayout>
```

4. Add a class file under the package (see Figure 3-17) and name it
Fragment1.java.

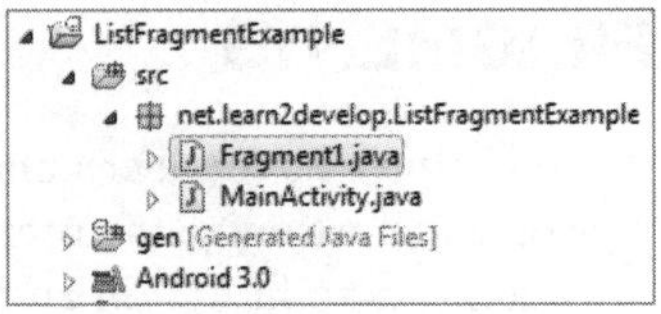

5. Populate the Fragment1.java file as follows:

FIGURE 3-17

```java
package net.learn2develop.ListFragmentExample;

import android.os.Bundle;
import android.view.LayoutInflater;
import android.view.View;
import android.view.ViewGroup;
import android.widget.ArrayAdapter;
import android.widget.ListView;
import android.widget.Toast;
import android.app.ListFragment;

public class Fragment1 extends ListFragment {
    String[] presidents = {
        "Dwight D. Eisenhower",
        "John F. Kennedy",
        "Lyndon B. Johnson",
        "Richard Nixon",
        "Gerald Ford",
        "Jimmy Carter",
        "Ronald Reagan",
        "George H. W. Bush",
        "Bill Clinton",
        "George W. Bush",
        "Barack Obama"
    };

    @Override
    public View onCreateView(LayoutInflater inflater,
    ViewGroup container, Bundle savedInstanceState) {
        return inflater.inflate(R.layout.fragment1, container, false);
    }

    @Override
    public void onCreate(Bundle savedInstanceState) {
        super.onCreate(savedInstanceState);
        setListAdapter(new ArrayAdapter<String>(getActivity(),
            android.R.layout.simple_list_item_1, presidents));
    }

    public void onListItemClick(ListView parent, View v,
    int position, long id)
    {
        Toast.makeText(getActivity(),
            "You have selected " + presidents[position],
            Toast.LENGTH_SHORT).show();
```

```
        }

    }
```

6. Modify the `main.xml` file as shown in bold:

```xml
<?xml version="1.0" encoding="utf-8"?>
<LinearLayout xmlns:android="http://schemas.android.com/apk/res/android"
    android:orientation="horizontal"
    android:layout_width="fill_parent"
    android:layout_height="fill_parent">
    <fragment
        android:name="net.learn2develop.ListFragmentExample.Fragment1"
        android:id="@+id/fragment1"
        android:layout_weight="1"
        android:layout_width="0dp"
        android:layout_height="match_parent" />
</LinearLayout>
```

7. Press F11 to debug the application on the Android emulator. Figure 3-18 shows the list fragment displaying the list of presidents' names.

FIGURE 3-18

8. Click on any of the items in the `ListView` and a message is displayed (see Figure 3-19).

FIGURE 3-19

How It Works

First, you created the XML file for the fragment by adding a `ListView` element to it:

```xml
<?xml version="1.0" encoding="utf-8"?>
<LinearLayout xmlns:android="http://schemas.android.com/apk/res/android"
    android:orientation="vertical"
    android:layout_width="fill_parent"
    android:layout_height="fill_parent">
    <ListView
        android:id="@id/android:list"
        android:layout_width="match_parent"
        android:layout_height="match_parent"
        android:layout_weight="1"
        android:drawSelectorOnTop="false"/>
</LinearLayout>
```

To create a list fragment, the Java class for the fragment must extend the `ListFragment` base class:

```java
public class Fragment1 extends ListFragment {

}
```

You then declared an array to contain the list of presidents' names:

```java
String[] presidents = {
    "Dwight D. Eisenhower",
    "John F. Kennedy",
    "Lyndon B. Johnson",
```

```
                "Richard Nixon",
                "Gerald Ford",
                "Jimmy Carter",
                "Ronald Reagan",
                "George H. W. Bush",
                "Bill Clinton",
                "George W. Bush",
                "Barack Obama"
        };
```

In the onCreate() event, you use the setListAdapter() method to programmatically fill the ListView with the content of the array. The ArrayAdapter object manages the array of strings that will be displayed by the ListView. In the preceding example, you set the ListView to display in the simple_list_item_1 mode:

```
        @Override
        public void onCreate(Bundle savedInstanceState) {
            super.onCreate(savedInstanceState);
            setListAdapter(new ArrayAdapter<String>(getActivity(),
                android.R.layout.simple_list_item_1, presidents));
        }
```

The onListItemClick() method is fired whenever an item in the ListView is clicked:

```
        public void onListItemClick(ListView parent, View v,
        int position, long id)
        {
            Toast.makeText(getActivity(),
                "You have selected " + presidents[position],
                Toast.LENGTH_SHORT).show();
        }
```

DialogFragment

Another type of fragment that you can create is the dialog fragment. A dialog fragment floats on top of an activity and is displayed modally. Dialog fragments are useful in cases where you need to obtain the user's response before continuing with the execution. To create a dialog fragment, you need to extend the DialogFragment base class.

The following Try It Out shows how to create a dialog fragment.

TRY IT OUT Creating and Using a Dialog Fragment

codefile DialogFragmentExample.zip available for download at Wrox.com

1. Using Eclipse, create an Android project and name it **DialogFragmentExample.**

2. Add a class file under the package and name it **Fragment1.java.**

3. Populate the `Fragment1.java` file as follows:

```java
package net.learn2develop.DialogFragmentExample;

import android.app.AlertDialog;
import android.app.Dialog;
import android.app.DialogFragment;
import android.content.DialogInterface;
import android.os.Bundle;

public class Fragment1 extends DialogFragment {

    static Fragment1 newInstance(String title) {
        Fragment1 fragment = new Fragment1();
        Bundle args = new Bundle();
        args.putString("title", title);
        fragment.setArguments(args);
        return fragment;
    }

    @Override
    public Dialog onCreateDialog(Bundle savedInstanceState) {
        String title = getArguments().getString("title");
        return new AlertDialog.Builder(getActivity())
            .setIcon(R.drawable.icon)
            .setTitle(title)
            .setPositiveButton("OK",
                new DialogInterface.OnClickListener() {
                    public void onClick(DialogInterface dialog,
                    int whichButton) {
                        ((MainActivity)getActivity()).doPositiveClick();
                    }
                }
            )
            .setNegativeButton("Cancel",
                new DialogInterface.OnClickListener() {
                    public void onClick(DialogInterface dialog,
                    int whichButton) {
                        ((MainActivity)getActivity()).doNegativeClick();
                    }
                }
            )
        .create();
    }

}
```

4. Populate the `MainActivity.java` file as shown here in bold:

```java
package net.learn2develop.DialogFragmentExample;

import android.app.Activity;
import android.os.Bundle;
```

```
import android.util.Log;

import android.app.FragmentManager;
import android.app.FragmentTransaction;

public class MainActivity extends Activity {
    /** Called when the activity is first created. */
    @Override
    public void onCreate(Bundle savedInstanceState) {
        super.onCreate(savedInstanceState);
        setContentView(R.layout.main);

        Fragment1 dialogFragment = Fragment1.newInstance(
            "Are you sure you want to do this?");
        dialogFragment.show(getFragmentManager(), "dialog");
    }

    public void doPositiveClick() {
        //---perform steps when user clicks on OK---
        Log.i("DialogFragmentExample", "User clicks on OK");
    }

    public void doNegativeClick() {
        //---perform steps when user clicks on Cancel---
        Log.i("DialogFragmentExample ", "User clicks on Cancel");
    }
}
```

5. Press F11 to debug the application on the Android emulator. Figure 3-20 shows the fragment displayed as an alert dialog.

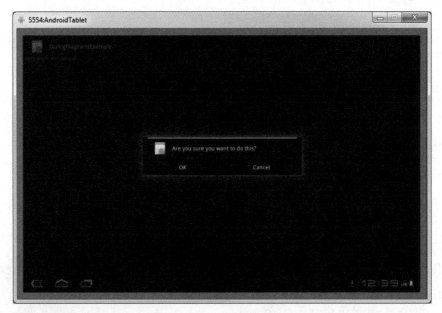

FIGURE 3-20

How It Works

To create a dialog fragment, first your Java class must extend the `DialogFragment` base class:

```
public class Fragment1 extends DialogFragment {

}
```

In this example, you created an alert dialog, which is a dialog window that displays a message with optional buttons. Within the `Fragment1` class, you defined the `newInstance()` method:

```
static Fragment1 newInstance(String title) {
    Fragment1 fragment = new Fragment1();
    Bundle args = new Bundle();
    args.putString("title", title);
    fragment.setArguments(args);
    return fragment;
}
```

The `newInstance()` method allows a new instance of the fragment to be created, and at the same time it accepts an argument specifying the string (`title`) to display in the alert dialog. The `title` is then stored in a `Bundle` object for use later.

Next, you defined the `onCreateDialog()` method, which is called after `onCreate()` and before `onCreateView()`:

```
@Override
public Dialog onCreateDialog(Bundle savedInstanceState) {
    String title = getArguments().getString("title");
    return new AlertDialog.Builder(getActivity())
        .setIcon(R.drawable.icon)
        .setTitle(title)
        .setPositiveButton("OK",
            new DialogInterface.OnClickListener() {
                public void onClick(DialogInterface dialog,
                int whichButton) {
                    ((MainActivity)getActivity()).doPositiveClick();
                }
            }
        )
        .setNegativeButton("Cancel",
            new DialogInterface.OnClickListener() {
                public void onClick(DialogInterface dialog,
                int whichButton) {
                    ((MainActivity)getActivity()).doNegativeClick();
                }
            }
        )
    .create();
}
```

Here, you created an alert dialog with two buttons: OK and Cancel. The string to be displayed in it is obtained from the `title` argument saved in the `Bundle` object.

To display the dialog fragment, you created an instance of it and then called its `show()` method:

```
Fragment1 dialogFragment = Fragment1.newInstance(
    "Are you sure you want to do this?");
dialogFragment.show(getFragmentManager(), "dialog");
```

You also needed to implement two methods, `doPositiveClick()` and `doNegativeClick()`, to handle the user clicking the OK and Cancel buttons, respectively:

```
public void doPositiveClick() {
    //---perform steps when user clicks on OK---
    Log.i("DialogFragmentExample", "User clicks on OK");
}

public void doNegativeClick() {
    //---perform steps when user clicks on Cancel---
    Log.i("DialogFragmentExample ", "User clicks on Cancel");
}
```

PreferenceFragment

Your Android applications will typically provide preferences that allow users to personalize the application for their own usage. For example, you may allow users to save the login credentials that they use to access their web resources, or save information such as how often the feeds must be refreshed (such as in a RSS reader application), and so on. In Android, you can use the `PreferenceActivity` base class to display an activity for the user to edit the preferences. In Android 3.0, you can now use the `PreferenceFragment` class to do the same thing.

The following Try It Out shows you how to create and use a preference fragment in Android 3.0.

TRY IT OUT   Creating and Using a Preference Fragment

codefile PreferenceFragmentExample.zip available for download at Wrox.com

1. Using Eclipse, create an Android project and name it **PreferenceFragmentExample.**

2. Create a new `xml` folder under the `res` folder and then add a new Android XML file to it. Name the XML file **preferences.xml** (see Figure 3-21).

3. Populate the `preferences.xml` file as follows:

```
<?xml version="1.0" encoding="utf-8"?>
<PreferenceScreen
    xmlns:android="http://schemas.android.com/apk/res/android">

    <PreferenceCategory android:title="Category 1">
        <CheckBoxPreference
```

FIGURE 3-21

```xml
                    android:title="Checkbox"
                    android:defaultValue="false"
                    android:summary="True of False"
                    android:key="checkboxPref" />
            </PreferenceCategory>

        <PreferenceCategory android:title="Category 2">
            <EditTextPreference
                    android:name="EditText"
                    android:summary="Enter a string"
                    android:defaultValue="[Enter a string here]"
                    android:title="Edit Text"
                    android:key="editTextPref" />
            <RingtonePreference
                    android:name="Ringtone Preference"
                    android:summary="Select a ringtone"
                    android:title="Ringtones"
                    android:key="ringtonePref" />
            <PreferenceScreen
                    android:title="Second Preference Screen"
                    android:summary=
                        "Click here to go to the second Preference Screen"
                    android:key="secondPrefScreenPref">
                <EditTextPreference
                        android:name="EditText"
                        android:summary="Enter a string"
                        android:title="Edit Text (second Screen)"
                        android:key="secondEditTextPref" />
            </PreferenceScreen>
        </PreferenceCategory>

</PreferenceScreen>
```

4. Add a class file under the package and name it **Fragment1.java**.

5. Populate the Fragment1.java file as follows:

```java
package net.learn2develop.PreferenceFragmentExample;

import android.os.Bundle;
import android.preference.PreferenceFragment;

public class Fragment1 extends PreferenceFragment {
    @Override
    public void onCreate(Bundle savedInstanceState) {
        super.onCreate(savedInstanceState);

        //--load the preferences from an XML file---
        addPreferencesFromResource(R.xml.preferences);
    }
}
```

6. Modify the `MainActivity.java` file as shown in bold:

```java
package net.learn2develop.PreferenceFragmentExample;

import android.app.Activity;
import android.os.Bundle;

import android.app.FragmentManager;
import android.app.FragmentTransaction;

public class MainActivity extends Activity {
    /** Called when the activity is first created. */
    @Override
    public void onCreate(Bundle savedInstanceState) {
        super.onCreate(savedInstanceState);
        setContentView(R.layout.main);

        FragmentManager fragmentManager = getFragmentManager();
        FragmentTransaction fragmentTransaction =
            fragmentManager.beginTransaction();
        Fragment1 fragment1 = new Fragment1();
        fragmentTransaction.replace(android.R.id.content, fragment1);
        fragmentTransaction.addToBackStack(null);
        fragmentTransaction.commit();
    }
}
```

7. Press F11 to debug the application on the Android emulator. Figure 3-22 shows the preference fragment displaying the list of preferences that the user can modify.

FIGURE 3-22

8. When the Edit Text preference is clicked, a popup will be displayed (see Figure 3-23).

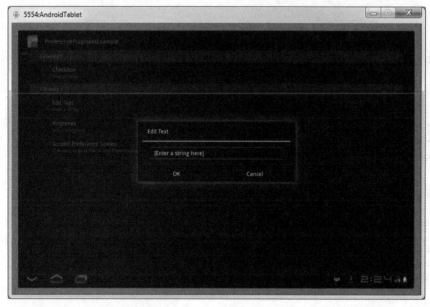

FIGURE 3-23

9. Clicking on the Second Preference Screen item will cause a second preference screen to be displayed (see Figure 3-24).

FIGURE 3-24

10. To cause the preference fragment to go away, click the Back button located in the lower-left corner of the screen.

11. If you look at the File Explorer (available in the DDMS perspective), you will be able to locate the preferences file located in the `/data/data/net.learn2develop.PreferenceFragmentExample/` `shared_prefs/` folder (see Figure 3-25). All the changes made by the user will be persisted in this file.

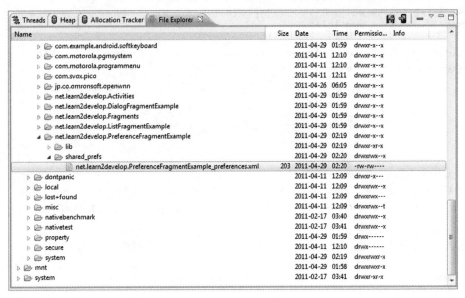

FIGURE 3-25

How It Works

To create a list of preferences in your Android application, you first needed to create the `preferences` `.xml` file and populate it with the various XML elements. This XML file defines the various items that you want to persist in your application.

To create the preference fragment, you needed to extend the `PreferenceFragment` base class:

```
public class Fragment1 extends PreferenceFragment {

}
```

To load the preferences file in the preference fragment, you use the `addPreferencesFromResource()` method:

```
@Override
public void onCreate(Bundle savedInstanceState) {
    super.onCreate(savedInstanceState);

    //--load the preferences from an XML file---
    addPreferencesFromResource(R.xml.preferences);
}
```

To display the preference fragment in your activity, you can make use of the `FragmentManager` and the `FragmentTransaction` classes:

```
FragmentManager fragmentManager = getFragmentManager();
FragmentTransaction fragmentTransaction =
    fragmentManager.beginTransaction();
Fragment1 fragment1 = new Fragment1();
fragmentTransaction.replace(android.R.id.content, fragment1);
fragmentTransaction.addToBackStack(null);
fragmentTransaction.commit();
```

You needed to add the preference fragment to the back stack using the `addToBackStack()` method so that the user can dismiss the fragment by clicking the Back button.

SUMMARY

In this chapter, you have learned how user interfaces are created in Android. You have also learned about the different layouts and views that you can use to build the UI in your Android application. Layouts views help to arrange the various views in the user interface of your Android application.

Finally, this chapter concluded with an overview of some of the specialized types of fragments that you can create in Android 3.0 for tablet applications. The three types of specialized fragments are `ListFragment` (for showing a list of items through a `ListView`), `DialogFragment` (for displaying as a dialog window), and `PreferenceFragment` (for displaying the shared preferences).

EXERCISES

1. What is the difference between the `dp` unit and the `px` unit? Which one should you use to specify the dimension of a view?

2. Why is the `AbsoluteLayout` not recommended for use?

3. How do you programmatically determine whether a `RadioButton` is checked?

4. Name the three specialized fragments you can use in your Android application.

Answers to the Exercises can be found in Appendix C.

▶ **WHAT YOU LEARNED IN THIS CHAPTER**

TOPIC	KEY CONCEPTS
LinearLayout	Arranges views in a single column or single row.
AbsoluteLayout	Enables you to specify the exact location of its children.
TableLayout	Groups views into rows and columns.
RelativeLayout	Enables you to specify how child views are positioned relative to each other.
FrameLayout	An on-screen placeholder that you can use to display a single view.
ScrollView	A special type of `FrameLayout` in that it enables users to scroll through a list of views that occupy more space than the physical display allows.
Unit of Measure	Use the `dp` to specify the dimension of views, and `sp` for font size.

TextView	`<TextView`
	    `android:layout_width="fill_parent"`
	    `android:layout_height="wrap_content"`
	    `android:text="@string/hello"`
	    `/>`

Button	`<Button android:id="@+id/btnSave"`
	    `android:layout_width="fill_parent"`
	    `android:layout_height="wrap_content"`
	    `android:text="Save" />`

ImageButton	`<ImageButton android:id="@+id/btnImg1"`
	    `android:layout_width="fill_parent"`
	    `android:layout_height="wrap_content"`
	    `android:src="@drawable/icon" />`

EditText	`<EditText android:id="@+id/txtName"`
	    `android:layout_width="fill_parent"`
	    `android:layout_height="wrap_content" />`

CheckBox	`<CheckBox android:id="@+id/chkAutosave"`
	    `android:layout_width="fill_parent"`
	    `android:layout_height="wrap_content"`
	    `android:text="Autosave" />`

continues

(continued)

TOPIC	KEY CONCEPTS
RadioGroup and RadioButton	```<RadioGroup android:id="@+id/rdbGp1" android:layout_width="fill_parent" android:layout_height="wrap_content" android:orientation="vertical" > <RadioButton android:id="@+id/rdb1" android:layout_width="fill_parent" android:layout_height="wrap_content" android:text="Option 1" /> <RadioButton android:id="@+id/rdb2" android:layout_width="fill_parent" android:layout_height="wrap_content" android:text="Option 2" /> </RadioGroup>```
ToggleButton	```<ToggleButton android:id="@+id/toggle1" android:layout_width="wrap_content" android:layout_height="wrap_content" />```
Specialized types of fragments	ListFragment, DialogFragment, and PreferenceFragment

PART II
Projects

Creating Location-Based Services Applications

➤ How to display Google Maps in your Android application

➤ How to display the zoom controls on the map

➤ Switching between the different map views

➤ Adding markers to maps

➤ How to get the address location touched on the map

➤ How to perform geocoding and reverse geocoding

➤ Obtaining geographical data using GPS, Cell-ID, and Wi-Fi triangulation

➤ How to monitor for a location

Everyone has seen the explosive growth of mobile apps in recent years. One category of apps that is very popular is *location-based services*, commonly known as LBS. LBS apps track your location, and may offer additional services such as locating amenities nearby, as well as offering suggestions for route planning, and so on. Of course, one of the key ingredients in an LBS app is maps, which present a visual representation of your location.

In this chapter, you will learn how to make use of Google Maps in your Android application, and how to manipulate it programmatically. In addition, you will learn how to obtain your geographical location using the `LocationManager` class available in the Android SDK. At the end of the chapter, you will have created a very cool Android tablet mapping application!

DISPLAYING MAPS

Google Maps is one of the many applications bundled with the Android platform. In addition to simply using the Maps application, you can also embed it into your own applications and make it do some very cool things. This section describes how to use Google Maps in your Android applications and programmatically perform the following:

➤   Change the views of Google Maps

➤   Obtain the latitude and longitude of locations in Google Maps

➤   Perform geocoding and reverse geocoding (translating an address to latitude and longitude and vice versa)

➤   Add markers to Google Maps

Creating the Project

To get started, you need to first create an Android project so you can display the Google Maps in your Android application.

TRY IT OUT   Creating the Project

codefile LBS.zip available for download at Wrox.com

1.   Using Eclipse, create an Android project as shown in Figure 4-1. Be sure to check the Google APIs checkbox in the Build Target section.

FIGURE 4-1

NOTE *In order to use Google Maps in your Android application, you need to ensure that you check the Google APIs as your build target. Google Maps is not part of the standard Android SDK, so you need to find it in the Google APIs add-on, as was discussed in Chapter 1.*

2. Once the project is created, observe the additional JAR file (maps .jar) located under the Google APIs folder (see Figure 4-2).

How It Works

This simple Try It Out created an Android project that uses the Google APIs add-on. The Google APIs add-on includes the standard Android library, with the addition of the Maps library, packaged within the maps.jar file.

FIGURE 4-2

Obtaining the Maps API Key

Beginning with the Android SDK release v1.0, you need to apply for a free Google Maps API key before you can integrate Google Maps into your Android application. When you apply for the key, you must also agree to Google's terms of use, so be sure to read them carefully.

To apply for a key, follow the series of steps outlined next.

 NOTE *Google provides detailed documentation on applying for a Maps API key at* http://code.google.com/android/add-ons/google-apis/mapkey.html.

First, if you are testing the application on the Android emulator or an Android device directly connected to your development machine, locate the SDK debug certificate located in the default folder (C:\Users\<username>\.android for Windows 7 users). You can verify the existence of the debug certificate by going to Eclipse and selecting Window ⇨ Preferences. Expand the Android item and select Build (see Figure 4-3). On the right side of the window, you will be able to see the debug certificate's location.

 NOTE *For Windows XP users, the default Android folder is* C: \Documents and Settings\<username>\Local Settings\Application Data\Android.

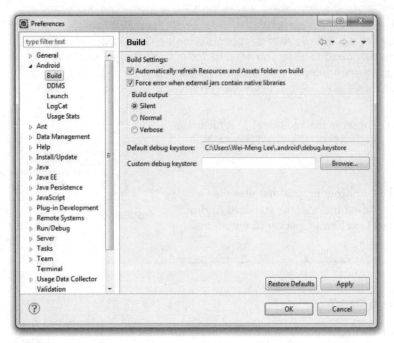

FIGURE 4-3

The filename of the debug keystore is debug.keystore. This is the certificate that Eclipse uses to sign your application so that it may be run on the Android emulator or other devices.

Using the debug keystore, you need to extract its MD5 fingerprint using the Keytool.exe application included with your JDK installation. This fingerprint is needed to apply for the free Google Maps key. You can usually find the Keytool.exe in the C:\Program Files\Java\<JDK_version_number>\bin folder.

Issue the following command (see Figure 4-4) to extract the MD5 fingerprint:

```
keytool.exe -list -alias androiddebugkey -keystore
"C:\Users\<username>\.android\debug.keystore" -storepass android
-keypass android
```

```
C:\Windows\system32\cmd.exe

C:\Program Files\Java\jre6\bin>keytool.exe -list -alias androiddebugkey -keystor
e "C:\Users\Wei-Meng Lee\.android\debug.keystore" -storepass android -keypass an
droid
androiddebugkey, Aug 3, 2010, PrivateKeyEntry,
Certificate fingerprint (MD5): EF:7A:61:EA:AF:E0:B4:2D:FD:43:5E:1D:26:04:34:BA

C:\Program Files\Java\jre6\bin>_
```

FIGURE 4-4

In this example, my MD5 fingerprint is EF:7A:61:EA:AF:E0:B4:2D:FD:43:5E:1D:26:04:34:BA.

Copy the MD5 certificate fingerprint and navigate your web browser to: http://code.google .com/android/maps-api-signup.html. Follow the instructions on the page to complete the application and obtain the Google Maps key. When you are done, you should see something similar to what is shown in Figure 4-5.

FIGURE 4-5

 NOTE *Although you can use the MD5 fingerprint of the debug keystore to obtain the Maps API key for debugging your application on the Android emulator or other devices, the key is not valid if you try to deploy your Android application as an APK file. When you are ready to deploy your application to the Android Market (or use another method of distribution), you need to reapply for a Maps API key using the certificate that will be used to sign your application. Chapter 6 discusses this topic in more detail.*

Displaying the Map

You are now ready to display Google Maps in your Android application. This involves two main tasks:

➤ Modify your AndroidManifest.xml file by adding both the <uses-library> element and the INTERNET permission.

➤ Add the MapView element to your UI.

The following Try It Out shows you how.

Displaying Google Maps

1. Using the project created in the previous section, add two XML files to the res/layout folder and name them **locations.xml** and **showmap.xml**. Also, add two class files under the package name and name them **Locations.java** and **ShowMap.java**. Figure 4-6 shows the four files added to the project.

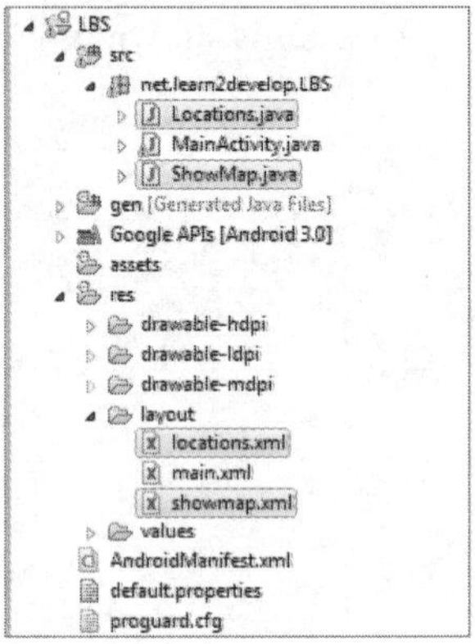

2. Populate locations.xml as follows:

```xml
<?xml version="1.0" encoding="utf-8"?>
<LinearLayout
    xmlns:android="http://schemas.android.com/apk/res/android"
    android:orientation="vertical"
    android:layout_width="fill_parent"
    android:layout_height="fill_parent">

<TextView
    android:layout_width="fill_parent"
    android:layout_height="wrap_content"
    android:text="List of Locations"
    android:textSize="30sp"
    android:textColor="#adff2f" />

<ListView
    android:id="@id/android:list"
    android:layout_width="match_parent"
    android:layout_height="match_parent"
    android:layout_weight="1"
    android:drawSelectorOnTop="false"/>

</LinearLayout>
```

FIGURE 4-6

3. Populate showmap.xml as follows (be sure to replace the value of the apiKey attribute with the API key you obtained earlier):

```xml
<?xml version="1.0" encoding="utf-8"?>
<LinearLayout
    xmlns:android="http://schemas.android.com/apk/res/android"
    android:orientation="vertical"
    android:layout_width="fill_parent"
    android:layout_height="fill_parent" >

<com.google.android.maps.MapView
    android:id="@+id/mapView"
    android:layout_width="fill_parent"
    android:layout_height="fill_parent"
    android:enabled="true"
    android:clickable="true"
```

```
                 android:apiKey="<your_maps_API_key_here>" />

</LinearLayout>
```

4. Populate the `Locations.java` file as follows:

```java
package net.learn2develop.LBS;

import android.app.ListFragment;
import android.os.Bundle;
import android.view.LayoutInflater;
import android.view.View;
import android.view.ViewGroup;
import android.widget.ArrayAdapter;
import android.widget.ListView;
import android.widget.Toast;

public class Locations extends ListFragment {
    String[] locations = {
        "Grand Canyon, Arizona (Valley)",
        "Bill Gates' house",
        "Yosemite National Park, California (Park)",
    };

    String[] latlng = {
        "36.1125,-113.995833",
        "47.627787,-122.242135",
        "36.849722,-119.5675",
    };

    @Override
    public void onCreate(Bundle savedInstanceState) {
        super.onCreate(savedInstanceState);
        //---displays the list of locations in the ListView---
        setListAdapter(new ArrayAdapter<String>(getActivity(),
            android.R.layout.simple_list_item_1, locations));
    }

    public void onListItemClick(ListView parent, View v,
    int position, long id)
    {
        Toast.makeText(getActivity(),
            "You have selected " + locations[position],
            Toast.LENGTH_SHORT).show();
    }

    @Override
    public View onCreateView(LayoutInflater inflater,
    ViewGroup container, Bundle savedInstanceState) {
        return inflater.inflate(R.layout.locations, container, false);
    }
}
```

5. Populate the ShowMap.java file as follows:

```java
package net.learn2develop.LBS;

import android.app.Fragment;
import android.os.Bundle;
import android.view.LayoutInflater;
import android.view.Menu;
import android.view.MenuInflater;
import android.view.MenuItem;
import android.view.View;
import android.view.ViewGroup;
import android.widget.Toast;

import com.google.android.maps.GeoPoint;
import com.google.android.maps.MapController;
import com.google.android.maps.MapView;

public class ShowMap extends Fragment {
    @Override
    public View onCreateView(LayoutInflater inflater,
    ViewGroup container, Bundle savedInstanceState) {
        return inflater.inflate(
            R.layout.showmap, container, false);
    }
}
```

6. Populate the main.xml file as follows:

```xml
<?xml version="1.0" encoding="utf-8"?>
<LinearLayout
    xmlns:android="http://schemas.android.com/apk/res/android"
    android:id="@+id/container"
    android:orientation="horizontal"
    android:layout_width="match_parent"
    android:layout_height="match_parent"
    >
    <fragment
        android:name="net.learn2develop.LBS.Locations"
        android:id="@+id/locationFragment"
        android:layout_weight="0.25"
        android:layout_width="20dp"
        android:layout_height="match_parent" />
    <fragment
        android:name="net.learn2develop.LBS.ShowMap"
        android:id="@+id/mapFragment"
        android:layout_weight="1"
        android:layout_width="0px"
        android:layout_height="match_parent" />

</LinearLayout>
```

7. Add the following statements in bold to the `MainActivity.java` file. Note that `MainActivity` is now extending the `MapActivity` class.

```
package net.learn2develop.LBS;

import android.os.Bundle;
import com.google.android.maps.MapActivity;

public class MainActivity extends MapActivity {

    /** Called when the activity is first created. */
    @Override
    public void onCreate(Bundle savedInstanceState) {
        super.onCreate(savedInstanceState);
        setContentView(R.layout.main);
    }

    @Override
    protected boolean isRouteDisplayed() {
        // TODO Auto-generated method stub
        return false;
    }
}
```

8. Add the following lines in bold to the `AndroidManifest.xml` file:

```
<?xml version="1.0" encoding="utf-8"?>
<manifest xmlns:android="http://schemas.android.com/apk/res/android"
    package="net.learn2develop.LBS"
    android:versionCode="1"
    android:versionName="1.0">
    <uses-sdk android:minSdkVersion="11" />

    <uses-permission android:name="android.permission.INTERNET">
    </uses-permission>

    <application android:icon="@drawable/icon" android:label="Where Am I">
        <uses-library android:name="com.google.android.maps" />
        <activity android:name=".MainActivity"
                android:label="@string/app_name">
            <intent-filter>
                <action android:name="android.intent.action.MAIN" />
                <category android:name="android.intent.category.LAUNCHER" />
            </intent-filter>
        </activity>
    </application>
</manifest>
```

9. Press F11 to debug the application on the Android emulator. Figure 4-7 shows the Android emulator displaying two fragments — one showing a list of locations and another showing Google Maps.

FIGURE 4-7

How It Works

In this Try It Out, you have two fragments, `Locations` and `ShowMap`. The `Locations` fragment is a list fragment showing a list of locations. The `ShowMap` fragment displays Google Maps. Note that any activity that is going to host a fragment that displays Google Maps must extend the `MapActivity` base class, which itself is an extension of the `Activity` class. Hence, you needed to make the following changes to `MainActivity.java`:

```
public class MainActivity extends MapActivity {

    /** Called when the activity is first created. */
    @Override
    public void onCreate(Bundle savedInstanceState) {
        super.onCreate(savedInstanceState);
        setContentView(R.layout.main);
    }

    @Override
    protected boolean isRouteDisplayed() {
        // TODO Auto-generated method stub
        return false;
    }
}
```

For the `MapActivity` base class, you need to implement one method: `isRouteDisplayed()`. This method is used for Google's accounting purposes, and you should return `true` for this method if you are displaying routing information on the map. For most simple cases, you can simply return `false`.

In order to display Google Maps in your application, you need to have the `INTERNET` permission in your manifest file. You then add the `<com.google.android.maps.MapView>` element to the XML file to embed the map within the fragment.

The `Locations` fragment contains a `TextView` showing the text "List of Locations" displayed in green. It also contains a `ListView` to display the list of locations:

```xml
<?xml version="1.0" encoding="utf-8"?>
<LinearLayout
    xmlns:android="http://schemas.android.com/apk/res/android"
    android:orientation="vertical"
    android:layout_width="fill_parent"
    android:layout_height="fill_parent">

<TextView
    android:layout_width="fill_parent"
    android:layout_height="wrap_content"
    android:text="List of Locations"
    android:textSize="30sp"
    android:textColor="#adff2f" />

<ListView
    android:id="@id/android:list"
    android:layout_width="match_parent"
    android:layout_height="match_parent"
    android:layout_weight="1"
    android:drawSelectorOnTop="false"/>

</LinearLayout>
```

The `Locations` class extends the `ListFragment` base class. It also contains two arrays — `locations` contains the names of various locations and `latlng` contains the corresponding latitude and longitude pair for each location stored in the `locations` array:

```java
public class Locations extends ListFragment {
    String[] locations = {
        "Grand Canyon, Arizona (Valley)",
        "Bill Gates' house",
        "Yosemite National Park, California (Park)",
    };

    String[] latlng = {
        "36.1125,-113.995833",
        "47.627787,-122.242135",
        "36.849722,-119.5675",
    };

    @Override
    public void onCreate(Bundle savedInstanceState) {
        super.onCreate(savedInstanceState);
        //---displays the list of locations in the ListView---
        setListAdapter(new ArrayAdapter<String>(getActivity(),
            android.R.layout.simple_list_item_1, locations));
    }

    public void onListItemClick(ListView parent, View v,
    int position, long id)
    {
```

```
            Toast.makeText(getActivity(),
                "You have selected " + locations[position],
                Toast.LENGTH_SHORT).show();
        }

        @Override
    public View onCreateView(LayoutInflater inflater,
    ViewGroup container, Bundle savedInstanceState) {
            return inflater.inflate(R.layout.locations, container, false);
        }
    }
```

When a location is clicked (or tapped on a real device), you display the name of the location using the
`Toast` class.

The `ShowMap` fragment simply loads the `showmap.xml` file to display Google Maps:

```
    public class ShowMap extends Fragment {
        @Override
        public View onCreateView(LayoutInflater inflater,
        ViewGroup container, Bundle savedInstanceState) {
            return inflater.inflate(
                R.layout.showmap, container, false);
        }
    }
```

CAN'T SEE THE MAP?

If instead of seeing Google Maps displayed you see an empty screen with grids,
then most likely you are using the wrong API key in the `main.xml` file. It is also
possible that you omitted the `INTERNET` permission in your `AndroidManifest.xml`
file. Finally, ensure that you have Internet access on your emulator/devices.

If your program does not run (i.e., it crashes), then you probably forgot to add the
following statement to the `AndroidManifest.xml` file:

```
        <uses-library android:name="com.google.android.maps" />
```

Note its placement in the `AndroidManifest.xml` file; it should be within the
`<Application>` element.

Displaying the Zoom Control

The previous section showed how you can display Google Maps in your Android application. You can
pan the map to any desired location and it will be updated on the fly. However, on the emulator there
is no way to zoom in or out from a particular location (on a real Android device you can pinch the
map to zoom it). Thus, in this section, you learn how you can enable users to zoom in or out of
the map using the built-in zoom controls.

TRY IT OUT Displaying the Built-In Zoom Controls

1. Using the project created in the previous activity, add the following statements in bold to the
   ShowMap.java file:

```java
public class ShowMap extends Fragment {
    private MapView mapView;

    @Override
    public View onCreateView(LayoutInflater inflater,
    ViewGroup container, Bundle savedInstanceState) {
        return inflater.inflate(
            R.layout.showmap, container, false);
    }

    @Override
    public void onStart() {
        super.onStart();
        mapView = (MapView) getActivity().findViewById(R.id.mapView);
        mapView.setBuiltInZoomControls(true);
    }
}
```

2. Press F11 to debug the application on the Android emulator. Observe the built-in zoom controls
   that appear at the bottom of the map when you click it (see Figure 4-8). Click the minus (–) icon
   to zoom out of the map, and the plus (+) icon to zoom into the map.

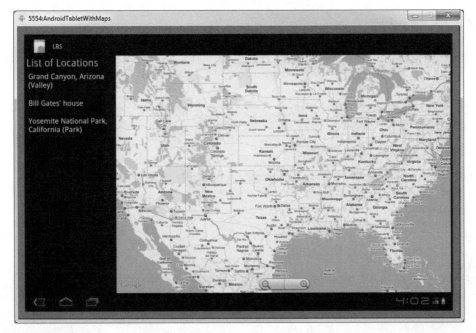

FIGURE 4-8

How It Works

To display the built-in zoom controls, you first get a reference to the map and then call the `setBuiltInZoomControls()` method:

```
mapView = (MapView) getActivity().findViewById(R.id.mapView);
mapView.setBuiltInZoomControls(true);
```

Besides displaying the zoom controls, you can also programmatically zoom in or out of the map by obtaining an instance of the `MapController` class from the `MapView` object and then calling the `zoomIn()` or `zoomOut()` method of the `MapController` class:

```
public class ShowMap extends Fragment {
    private MapView mapView;
    private MapController mc;

    @Override
    public View onCreateView(LayoutInflater inflater,
    ViewGroup container, Bundle savedInstanceState) {
        return inflater.inflate(
            R.layout.showmap, container, false);
    }

    @Override
    public void onStart() {
        super.onStart();
        mapView = (MapView) getActivity().findViewById(R.id.mapView);
        mapView.setBuiltInZoomControls(true);

        mc = mapView.getController();
        mc.zoomIn();  //---zoom into the map---
        mc.zoomOut(); //---zoom out of the map---
    }
}
```

Changing Views

By default, Google Maps is displayed in a map view, which is basically drawings of streets and places of interest. You can also set Google Maps to display in satellite view using the `setSatellite()` method of the `MapView` class:

```
@Override
public void onStart() {
    super.onStart();
    mapView = (MapView) getActivity().findViewById(R.id.mapView);
```

```
        mapView.setBuiltInZoomControls(true);

        mc = mapView.getController();
        mapView.setSatellite(true);
    }
```

Figure 4-9 shows Google Maps displayed in satellite view.

FIGURE 4-9

If you want to display traffic conditions on the map, use the setTraffic() method:

```
        mapView.setTraffic(true);
```

Figure 4-10 shows the map displaying the current traffic conditions. The different colors reflect the varying traffic conditions. In general, green means smooth traffic of about 50 miles per hour, yellow means moderate traffic of about 25–50 miles per hour, and red means slow traffic of about less than 25 miles per hour.

Currently, traffic information is available only in major cities in the United States, France, Britain, Australia, and Canada, although new cities and countries are frequently added.

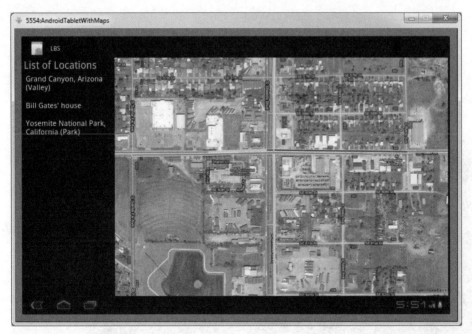

FIGURE 4-10

Navigating to a Specific Location

By default, Google Maps displays the map of the United States when it is first loaded. However, you can set Google Maps to display a particular location. To do so, use the `animateTo()` method of the `MapController` class.

The following Try It Out shows how you can programmatically animate Google Maps to a particular location.

TRY IT OUT   Setting the Map to Display a Specific Location

1. Using the project created in the previous activity, add the following statements in bold to the `ShowMap.java` file:

```
public class ShowMap extends Fragment {
    private MapView mapView;
    private MapController mc;
    private GeoPoint p;

    @Override
    public View onCreateView(LayoutInflater inflater,
    ViewGroup container, Bundle savedInstanceState) {
        return inflater.inflate(
```

```
                    R.layout.showmap, container, false);
    }

    @Override
    public void onStart() {
        super.onStart();
        mapView = (MapView) getActivity().findViewById(R.id.mapView);
        mc = mapView.getController();

        mapView.setBuiltInZoomControls(true);
        mapView.setSatellite(true);
        mapView.setTraffic(true);
    }

    //---go to a particular location---
    public void gotoLocation(String latlng)
    {
        //---the location is represented as "lat,lng"---
        String[] coordinates = latlng.split(",");
        double lat = Double.parseDouble(coordinates[0]);
        double lng = Double.parseDouble(coordinates[1]);
        p = new GeoPoint((int) (lat * 1E6),
                         (int) (lng * 1E6));
        mc.animateTo(p);
        mc.setZoom(16);
        mapView.invalidate();
    }
}
```

2. Add the following statements in bold to the Locations.java file:

```
public void onListItemClick(ListView parent, View v,
int position, long id)
{
    Toast.makeText(getActivity(),
        "You have selected " + locations[position],
        Toast.LENGTH_SHORT).show();

    //---obtain a reference to the ShowMap fragment---
    ShowMap mapFragment =
        (ShowMap)getFragmentManager().findFragmentById(
          R.id.mapFragment);
    //---invoke the method from the fragment---
    mapFragment.gotoLocation(latlng[position]);
}
```

3. Press F11 to debug the application on the Android emulator. Click a particular location listed in the left fragment. Observe that the map animates to the selected location (see Figure 4-11).

FIGURE 4-11

How It Works

In the `ShowMap` fragment, you defined the `gotoLocation()` method, which takes a single input string argument containing the location in the following format: `Latitude,longitude`. From this string, you can extract the latitude and longitude of the location and then use a `GeoPoint` object to represent a geographical location. Note that for this class, the latitude and longitude of a location are represented in micro degrees. This means that they are stored as integer values. For a latitude value of 40.747778, for example, you need to multiply it by 1e6 (which is one million) to obtain 40747778.

To navigate the map to a particular location, you used the `animateTo()` method of the `MapController` class. The `setZoom()` method enables you to specify the zoom level at which the map is displayed (the bigger the number, the more details you see on the map). The `invalidate()` method forces the `MapView` to be redrawn.

In the `Locations` fragment, when a location is clicked in the `ListView`, you call the `gotoLocation()` method that you have defined in the `ShowMap` fragment:

```
//---obtain a reference to the ShowMap fragment---
ShowMap mapFragment =
    (ShowMap)getFragmentManager().findFragmentById(
    R.id.mapFragment);
//---invoke the method from the fragment---
mapFragment.gotoLocation(latlng[position]);
```

You made use of the `getFragmentManager()` method to obtain an instance of the `FragmentManager` object, and then called its `findFragmentById()` method to obtain an instance of the `ShowMap` fragment. You then called the `gotoLocation()` method available in the fragment.

Adding Markers

Adding markers to a map to indicate places of interest enables your users to easily locate the places they are looking for. The following Try It Out shows you how to add a marker to Google Maps.

TRY IT OUT   Adding Markers to the Map

1. Create a GIF image containing a pushpin (see Figure 4-12) and copy it into the `res/drawable-mdpi` folder of the project. For the best effect, make the background of the image transparent so that it does not block parts of the map when the image is added to the map.

2. Using the project created in the previous activity, add the following statements in bold to the `ShowMap.java` file:

FIGURE 4-12

```java
package net.learn2develop.LBS;

import com.google.android.maps.GeoPoint;
import com.google.android.maps.MapController;
import com.google.android.maps.MapView;

//...

import android.graphics.Bitmap;
import android.graphics.BitmapFactory;
import android.graphics.Canvas;
import android.graphics.Point;
import com.google.android.maps.Overlay;
import java.util.List;

public class ShowMap extends Fragment {
    private MapView mapView;
    private MapController mc;
    private GeoPoint p;

    class MapOverlay extends com.google.android.maps.Overlay
    {
        @Override
        public boolean draw(Canvas canvas, MapView mapView,
        boolean shadow, long when)
        {
            super.draw(canvas, mapView, shadow);

            //---translate the GeoPoint to screen pixels---
            Point screenPts = new Point();
```

```
            mapView.getProjection().toPixels(p, screenPts);

            //---add the marker---
            Bitmap bmp = BitmapFactory.decodeResource(
                getResources(), R.drawable.pushpin);
            canvas.drawBitmap(bmp, screenPts.x, screenPts.y-50, null);
            return true;
        }
    }

    @Override
    public View onCreateView(LayoutInflater inflater,
    ViewGroup container, Bundle savedInstanceState) {
        return inflater.inflate(
            R.layout.showmap, container, false);
    }

    @Override
    public void onStart() {
        super.onStart();
        mapView = (MapView) getActivity().findViewById(R.id.mapView);
        mc = mapView.getController();

        mapView.setBuiltInZoomControls(true);
        mapView.setSatellite(true);
        mapView.setTraffic(true);
    }

    //---go to a particular location---
    public void gotoLocation(String latlng)
    {
        //---the location is represented as "lat,lng"---
        String[] coordinates = latlng.split(",");
        double lat = Double.parseDouble(coordinates[0]);
        double lng = Double.parseDouble(coordinates[1]);
        p = new GeoPoint((int) (lat * 1E6),
                         (int) (lng * 1E6));
        mc.animateTo(p);
        mc.setZoom(16);

        //---Add a location marker---
        MapOverlay mapOverlay = new MapOverlay();
        List<Overlay> listOfOverlays = mapView.getOverlays();
        listOfOverlays.clear();
        listOfOverlays.add(mapOverlay);

        mapView.invalidate();
    }
}
```

3. Press F11 to debug the application on the Android emulator. Click a location on the left fragment
    to see the marker added to the map, as shown in Figure 4-13.

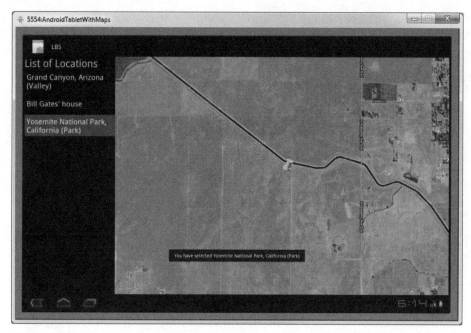

FIGURE 4-13

How It Works

To add a marker to the map, you first need to define a class that extends the `Overlay` class:

```
class MapOverlay extends com.google.android.maps.Overlay
{
    @Override
    public boolean draw(Canvas canvas, MapView mapView,
    boolean shadow, long when)
    {
        super.draw(canvas, mapView, shadow);

        //...
    }
}
```

An overlay represents an individual item that you can draw on the map. You can add as many overlays as you want. In the `MapOverlay` class, you override the `draw()` method so that you can draw the pushpin image on the map. In particular, note that you need to translate the geographical location (represented by a `GeoPoint` object, `p`) into screen coordinates:

```
//---translate the GeoPoint to screen pixels---
Point screenPts = new Point();
mapView.getProjection().toPixels(p, screenPts);
```

Because you want the pointed tip of the pushpin to indicate the position of the location, you need to deduct the height of the image (which is 50 pixels) from the y coordinate of the point (see Figure 4-14) and draw the image at that location:

```
//---add the marker---
Bitmap bmp = BitmapFactory.decodeResource(
    getResources(), R.drawable.pushpin);
canvas.drawBitmap(bmp, screenPts.x, screenPts.y-50, null);
```

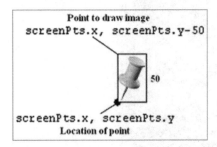

FIGURE 4-14

To add the marker, you create an instance of the MapOverlay class and add it to the list of overlays available on the MapView object:

```
//---Add a location marker---
MapOverlay mapOverlay = new MapOverlay();
List<Overlay> listOfOverlays = mapView.getOverlays();
listOfOverlays.clear();
listOfOverlays.add(mapOverlay);
```

Getting the Location That Was Touched

After using Google Maps for a while, you may want to know the latitude and longitude of a location corresponding to the position on the screen that was just touched. Knowing this information is very useful, as you can determine a location's address, a process known as *reverse geocoding* (you will learn how this is done in the next section).

If you have added an overlay to the map, you can override the onTouchEvent() method within the MapOverlay class. This method is fired every time the user touches the map. This method has two parameters: MotionEvent and MapView. Using the MotionEvent parameter, you can determine whether the user has lifted his or her finger from the screen using the getAction() method. In the following code snippet, if the user has touched and then lifted the finger, you display the latitude and longitude of the location touched:

```
import android.view.MotionEvent;

//...

class MapOverlay extends com.google.android.maps.Overlay
{
    @Override
    public boolean draw(Canvas canvas, MapView mapView,
    boolean shadow, long when)
    {
        super.draw(canvas, mapView, shadow);

        //---translate the GeoPoint to screen pixels---
        Point screenPts = new Point();
        mapView.getProjection().toPixels(p, screenPts);

        //---add the marker---
        Bitmap bmp = BitmapFactory.decodeResource(
            getResources(), R.drawable.pushpin);
        canvas.drawBitmap(bmp, screenPts.x, screenPts.y-50, null);
        return true;
    }

    @Override
    public boolean onTouchEvent(MotionEvent event, MapView mapView)
    {
        //---when user lifts his finger---
        if (event.getAction() == 1) {
            GeoPoint p = mapView.getProjection().fromPixels(
                (int) event.getX(),
                (int) event.getY());

            Toast.makeText(getActivity(),
                "Location: "+
                p.getLatitudeE6() / 1E6 + "," +
                p.getLongitudeE6() /1E6 ,
                Toast.LENGTH_SHORT).show();
        }
        return false;
    }
}
```

The getProjection() method returns a projection for converting between screen-pixel coordinates and latitude/longitude coordinates. The fromPixels() method then converts the screen coordinates into a GeoPoint object.

Figure 4-15 shows the map displaying a set of coordinates when the user clicks a location on the map.

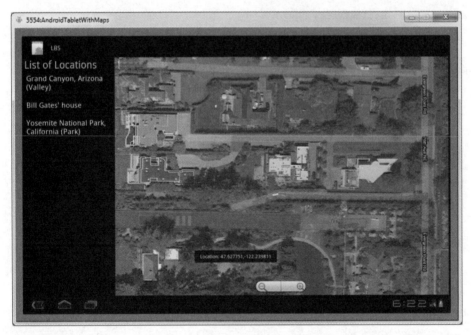

FIGURE 4-15

Geocoding and Reverse Geocoding

As mentioned in the preceding section, if you know the latitude and longitude of a location, you can find out its address using a process known as *reverse geocoding*. Google Maps in Android supports this via the Geocoder class. The following code snippet shows how you can retrieve the address of a location just touched using the getFromLocation() method:

```
import android.location.Address;
import android.location.Geocoder;
import java.util.Locale;
import java.io.IOException;
//...

    class MapOverlay extends com.google.android.maps.Overlay
    {
        @Override
        public boolean draw(Canvas canvas, MapView mapView,
        boolean shadow, long when)
        {
            super.draw(canvas, mapView, shadow);

            //---translate the GeoPoint to screen pixels---
            Point screenPts = new Point();
            mapView.getProjection().toPixels(p, screenPts);

            //---add the marker---
```

```java
        Bitmap bmp = BitmapFactory.decodeResource(
            getResources(), R.drawable.pushpin);
        canvas.drawBitmap(bmp, screenPts.x, screenPts.y-50, null);
        return true;
    }

    @Override
    public boolean onTouchEvent(MotionEvent event, MapView mapView)
    {
        //---when user lifts his finger---
        if (event.getAction() == 1) {
            GeoPoint p = mapView.getProjection().fromPixels(
                (int) event.getX(),
                (int) event.getY());

            /*
            Toast.makeText(getActivity(),
                "Location: "+
                p.getLatitudeE6() / 1E6 + "," +
                p.getLongitudeE6() /1E6 ,
                Toast.LENGTH_SHORT).show();
            */

            Geocoder geoCoder = new Geocoder(
                getActivity(), Locale.getDefault());
            try {
                List<Address> addresses = geoCoder.getFromLocation(
                    p.getLatitudeE6()  / 1E6,
                    p.getLongitudeE6() / 1E6, 1);

                String add = "";
                if (addresses.size() > 0) {
                    for (int i=0;
                        i<addresses.get(0).getMaxAddressLineIndex();
                        i++)
                        add += addresses.get(0).getAddressLine(i) +
                        "\n";
                }
                Toast.makeText(getActivity(), add,
                    Toast.LENGTH_SHORT).show();
            }
            catch (IOException e) {
                e.printStackTrace();
            }
            return true;
        }
        return false;
    }
}
```

The Geocoder object converts the latitude and longitude into an address using the getFromLocation() method. Once the address is obtained, you display it using the Toast class. Figure 4-16 shows the application displaying the address of a location that was touched on the map.

FIGURE 4-16

If you know the address of a location but want to know its latitude and longitude, you can do so via geocoding. Again, you can use the Geocoder class for this purpose. The following code shows how you can find the exact location of the Empire State Building by using the getFromLocationName() method:

```
//---geo-coding---
Geocoder geoCoder = new Geocoder(this, Locale.getDefault());
try {
    List<Address> addresses = geoCoder.getFromLocationName(
        "empire state building", 5);

    String add = "";
    if (addresses.size() > 0) {
        p = new GeoPoint(
                (int) (addresses.get(0).getLatitude() * 1E6),
                (int) (addresses.get(0).getLongitude() * 1E6));
        mc.animateTo(p);
        mapView.invalidate();
    }
} catch (IOException e) {
    e.printStackTrace();
}
```

GETTING LOCATION DATA

Nowadays, mobile devices are commonly equipped with GPS receivers. Because of the many satellites orbiting the earth, you can use a GPS receiver to find your location easily. However, GPS requires a clear sky to work and hence does not always work indoors or where satellites can't penetrate (such as a tunnel through a mountain).

Another effective way to locate your position is through *cell tower triangulation*. When a mobile phone is switched on, it is constantly in contact with base stations surrounding it. By knowing the identity of cell towers, it is possible to translate this information into a physical location through the use of various databases containing the cell towers' identities and their exact geographical locations. The advantage of cell tower triangulation is that it works indoors, without the need to obtain information from satellites. However, it is not as precise as GPS because its accuracy depends on overlapping signal coverage, which varies quite a bit. Cell tower triangulation works best in densely populated areas where the cell towers are closely located.

A third method of locating your position is to rely on Wi-Fi triangulation. Rather than connect to cell towers, the device connects to a Wi-Fi network and checks the service provider against databases to determine the location serviced by the provider. Of the three methods described here, Wi-Fi triangulation is the least accurate.

On the Android, the SDK provides the `LocationManager` class to help your device determine the user's physical location. The following Try It Out shows you how this is done in code.

TRY IT OUT | **Setting the Map to Display a Specific Location Using the Location Manager Class**

1. Using the same project created in the previous section, add the following statements in bold to the `MainActivity.java` file:

```
package net.learn2develop.LBS;

import com.google.android.maps.MapActivity;
import android.os.Bundle;
import android.widget.Toast;

import android.content.Context;
import android.location.Location;
import android.location.LocationListener;
import android.location.LocationManager;

public class MainActivity extends MapActivity {
    private LocationManager lm;
    private LocationListener locationListener;

    /** Called when the activity is first created. */
```

```java
@Override
public void onCreate(Bundle savedInstanceState) {
    super.onCreate(savedInstanceState);
    setContentView(R.layout.main);

    //---use the LocationManager class to obtain locations data---
    lm = (LocationManager)
        getSystemService(Context.LOCATION_SERVICE);
    locationListener = new MyLocationListener();
}

public void TrackingUsingGPS(boolean StartTracking)
{
    if (StartTracking) {
        lm.requestLocationUpdates(
            LocationManager.GPS_PROVIDER,
            0,
            0,
            locationListener);
    } else    {
        lm.removeUpdates(locationListener);
    }
}

public void TrackingUsingCellularWiFi(boolean StartTracking)
{
    if (StartTracking) {
        lm.requestLocationUpdates(
            LocationManager.NETWORK_PROVIDER,
            0,
            0,
            locationListener);
    } else    {
        lm.removeUpdates(locationListener);
    }
}

private class MyLocationListener implements LocationListener
{
    @Override
    public void onLocationChanged(Location loc) {
        if (loc != null) {
            Toast.makeText(getBaseContext(),
                "Location changed : Lat: " + loc.getLatitude() +
                " Lng: " + loc.getLongitude(),
                Toast.LENGTH_SHORT).show();
        }
        //---obtain a reference to the ShowMap fragment---
        ShowMap mapFragment =
            (ShowMap)getFragmentManager().findFragmentById(
            R.id.mapFragment);
        //---invoke the method from the fragment---
        mapFragment.gotoLocation(loc.getLatitude() +
```

```
                    "," + loc.getLongitude());
        }

        @Override
        public void onProviderDisabled(String provider) {
        }

        @Override
        public void onProviderEnabled(String provider) {
        }

        @Override
        public void onStatusChanged(String provider, int status,
            Bundle extras) {
        }
    }

    @Override
    protected boolean isRouteDisplayed() {
        // TODO Auto-generated method stub
        return false;
    }
}
```

2. Add the following statements in bold to the ShowMap.java file:

```
package net.learn2develop.LBS;

import com.google.android.maps.GeoPoint;
import com.google.android.maps.MapController;
import com.google.android.maps.MapView;
//...

import android.view.Menu;
import android.view.MenuInflater;
import android.view.MenuItem;

public class ShowMap extends Fragment {
    private MapView mapView;
    private MapController mc;
    private GeoPoint p;

    class MapOverlay extends com.google.android.maps.Overlay
    {
        //...
    }

    @Override
    public View onCreateView(LayoutInflater inflater,
    ViewGroup container, Bundle savedInstanceState) {
        return inflater.inflate(
            R.layout.showmap, container, false);
```

```java
    }

    @Override
    public void onStart() {
        super.onStart();
        mapView = (MapView) getActivity().findViewById(R.id.mapView);
        mc = mapView.getController();

        mapView.setBuiltInZoomControls(true);
        mapView.setSatellite(true);
        mapView.setTraffic(true);
    }

    //---go to a particular location---
    public void gotoLocation(String latlng)
    {
        //---the location is represented as "lat,lng"---
        String[] coordinates = latlng.split(",");
        double lat = Double.parseDouble(coordinates[0]);
        double lng = Double.parseDouble(coordinates[1]);
        p = new GeoPoint((int) (lat * 1E6),
                         (int) (lng * 1E6));
        mc.animateTo(p);
        mc.setZoom(16);

        //---Add a location marker---
        MapOverlay mapOverlay = new MapOverlay();
        List<Overlay> listOfOverlays = mapView.getOverlays();
        listOfOverlays.clear();
        listOfOverlays.add(mapOverlay);

        mapView.invalidate();
    }

    @Override
    public void onCreate (Bundle savedInstanceState) {
        super.onCreate(savedInstanceState);
        //---need to call this in order to fire the
        // onCreateOptionsMenu() event---
        setHasOptionsMenu(true);
    }

    //---creating action items on the action bar for a fragment---
    @Override
    public void onCreateOptionsMenu(Menu menu, MenuInflater inflater) {
        super.onCreateOptionsMenu(menu, inflater);
        CreateMenu(menu);
    }

    //---when a menu item is selected---
    @Override
    public boolean onOptionsItemSelected(MenuItem item)
```

```
{
    return MenuChoice(item);
}

//---create the action items---
private void CreateMenu(Menu menu)
{
    MenuItem mnu1 = menu.add(0, 0, 0, "Start Tracking");
    {
        mnu1.setShowAsAction(
            MenuItem.SHOW_AS_ACTION_IF_ROOM |
            MenuItem.SHOW_AS_ACTION_WITH_TEXT);
    }

    MenuItem mnu2 = menu.add(0, 1, 1, "Stop Tracking");
    {
        mnu2.setShowAsAction(
            MenuItem.SHOW_AS_ACTION_IF_ROOM |
            MenuItem.SHOW_AS_ACTION_WITH_TEXT);
    }

    MenuItem mnu3 = menu.add(0, 2, 1, "Use GPS");
    {
        mnu3.setShowAsAction(
            MenuItem.SHOW_AS_ACTION_WITH_TEXT);
        mnu3.setCheckable(true);
    }

    MenuItem mnu4 = menu.add(0, 3, 1, "Use Cellular/WiFi");
    {
        mnu4.setShowAsAction(
            MenuItem.SHOW_AS_ACTION_WITH_TEXT);
        mnu4.setCheckable(true);
    }
}

private boolean MenuChoice(MenuItem item)
{
    //---obtain an instance of the activity---
    MainActivity mainActivity;
    mainActivity = (MainActivity) getActivity();

    switch (item.getItemId()) {
    case 0:
        Toast.makeText(getActivity(), "Tracking turned on",
            Toast.LENGTH_LONG).show();
        //---calling the methods from the activity---
        mainActivity.TrackingUsingGPS(true);
        mainActivity.TrackingUsingCellularWiFi(true);
        return true;
    case 1:
        Toast.makeText(getActivity(), "Tracking turned off",
            Toast.LENGTH_LONG).show();
```

```
                mainActivity.TrackingUsingGPS(false);
                mainActivity.TrackingUsingCellularWiFi(false);
                return true;
        case 2:
            if (!item.isChecked()) {
                Toast.makeText(getActivity(),
                    "Using GPS for location tracking",
                    Toast.LENGTH_LONG).show();
            }
            item.setChecked(!(item.isChecked()));
            mainActivity.TrackingUsingGPS(item.isChecked());
            return true;
        case 3:
            if (!item.isChecked()) {
                Toast.makeText(getActivity(),
                    "Using Cellular/WiFi for location tracking",
                    Toast.LENGTH_LONG).show();
            }
            item.setChecked(!(item.isChecked()));
            mainActivity.TrackingUsingCellularWiFi(item.isChecked());
            return true;
        }
        return false;
    }
}
```

3. Add the following statements in bold to the AndroidManifest.xml file:

```
<?xml version="1.0" encoding="utf-8"?>
<manifest xmlns:android="http://schemas.android.com/apk/res/android"
    package="net.learn2develop.LBS"
    android:versionCode="1"
    android:versionName="1.0">
    <uses-sdk android:minSdkVersion="11" />

    <uses-permission android:name="android.permission.INTERNET">
    </uses-permission>
    <uses-permission
        android:name="android.permission.ACCESS_COARSE_LOCATION">
    </uses-permission>
    <uses-permission
        android:name="android.permission.ACCESS_FINE_LOCATION">
    </uses-permission>

    <application android:icon="@drawable/icon" android:label="Where Am I">
        <uses-library android:name="com.google.android.maps" />
        <activity android:name=".MainActivity"
                android:label="@string/app_name">
            <intent-filter>
                <action android:name="android.intent.action.MAIN" />
                <category android:name="android.intent.category.LAUNCHER" />
            </intent-filter>
        </activity>
    </application>
</manifest>
```

4. Press F11 to debug the application on the Android emulator. Figure 4-17 shows the action items displayed on the action bar.

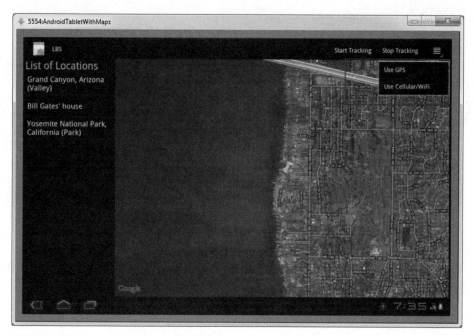

FIGURE 4-17

5. Clicking the Start Tracking item makes the application listen for location information using GPS, the cellular network, and Wi-Fi. Clicking the Stop Tracking item stops the application from listening for location information. Alternatively, you can also click the Use GPS item to only use GPS, and click Use Cellular/Wi-Fi to use cellular and Wi-Fi networks. For this exercise, click the Use GPS item.

 NOTE *When testing on the Android emulator, clicking the Start Tracking or Use Cellular/Wi-Fi items causes the application to crash. This is because the* NETWORK_PROVIDER *is not supported on the Android 3.0 emulator.*

6. To simulate GPS data received by the Android emulator, use the Location Controls tool (see Figure 4-18) located in the DDMS perspective.

FIGURE 4-18

7. Ensure that you have first selected the emulator in the Devices tab. Then, in the Emulator Control tab, locate the Location Controls tool and select the Manual tab. Enter a latitude and longitude and click the Send button.

8. Note that the map on the emulator now animates to another location (see Figure 4-19). This proves that the application has received the GPS data.

FIGURE 4-19

How It Works

In Android, location-based services are provided by the `LocationManager` class, located in the `android.location` package. Using the `LocationManager` class, your application can obtain periodic updates of the device's geographical locations, as well as fire an intent when it enters the proximity of a certain location.

In the `MainActivity.java` file, you first obtain a reference to the `LocationManager` class using the `getSystemService()` method. To be notified whenever there is a change in location using GPS, you need to register a request for location changes so that your program can be notified periodically. This is done via the `requestLocationUpdates()` method:

```
lm.requestLocationUpdates(
        LocationManager.GPS_PROVIDER,
        0,
        0,
        locationListener);
```

This method takes four parameters:

➤ `provider` — The name of the provider with which you register. In this case, you are using GPS to obtain your geographical location data.

➤ `minTime` — The minimum time interval for notifications, in milliseconds.

➤ `minDistance` — The minimum distance interval for notifications, in meters.

➤ `listener` — An object whose `onLocationChanged()` method will be called for each location update.

The `MyLocationListener` class implements the `LocationListener` abstract class. You overrode four methods in this implementation:

➤ `onLocationChanged(Location location)` — Called when the location has changed

➤ `onProviderDisabled(String provider)` — Called when the provider is disabled by the user

➤ `onProviderEnabled(String provider)` — Called when the provider is enabled by the user

➤ `onStatusChanged(String provider, int status, Bundle extras)` — Called when the provider status changes

In this example, you're more interested in what happens when a location changes, so you write some code in the `onLocationChanged()` method. Specifically, when a location changes, you display a small dialog on the screen showing the new location information: latitude and longitude. You show this dialog using the `Toast` class.

To use the Cellular and Wi-Fi networks (important for indoor use) to obtain your location data, you use the network location provider, like this:

```
lm.requestLocationUpdates(
        LocationManager.NETWORK_PROVIDER,
        0,
        0,
        locationListener);
```

You combine both the GPS location provider with the network location provider within your application.

To stop listening for location updates, you use the `removeUpdates()` method:

```
lm.removeUpdates(locationListener);
```

In the `ShowMap` fragment, you added four action items to the Action Bar, two of which are displayed on the Action Bar, and two under the overflow action item. In Chapter 2, you learned how to add action items to the Action Bar by overriding the `onCreateOptionsMenu()` method within the activity. However, in this case, the fragment is adding action items, instead of the activity. To enable a fragment to add action items, you need to call the `setHasOptionsMenu()` method so that the fragment fires the `onCreateOptionsMenu()`:

```
@Override
public void onCreate (Bundle savedInstanceState) {
    super.onCreate(savedInstanceState);
    //---need to call this in order to fire the
    // onCreateOptionsMenu() event---
    setHasOptionsMenu(true);
}
```

Once this is done, the rest is relatively straightforward:

```
//---creating action items on the action bar for a fragment---
@Override
public void onCreateOptionsMenu(Menu menu, MenuInflater
inflater) {
        super.onCreateOptionsMenu(menu, inflater);
        CreateMenu(menu);
}

//---when a menu item is selected---
@Override
public boolean onOptionsItemSelected(MenuItem item)
{
    return MenuChoice(item);
}

//---create the action items---
private void CreateMenu(Menu menu)
{
    MenuItem mnu1 = menu.add(0, 0, 0, "Start Tracking");
    {
        mnu1.setShowAsAction(
            MenuItem.SHOW_AS_ACTION_IF_ROOM |
            MenuItem.SHOW_AS_ACTION_WITH_TEXT);
    }

    MenuItem mnu2 = menu.add(0, 1, 1, "Stop Tracking");
    {
        mnu2.setShowAsAction(
            MenuItem.SHOW_AS_ACTION_IF_ROOM |
            MenuItem.SHOW_AS_ACTION_WITH_TEXT);
```

```
        }

    MenuItem mnu3 = menu.add(0, 2, 1, "Use GPS");
    {
        mnu3.setShowAsAction(
            MenuItem.SHOW_AS_ACTION_WITH_TEXT);
        mnu3.setCheckable(true);
    }

    MenuItem mnu4 = menu.add(0, 3, 1, "Use Cellular/WiFi");
    {
        mnu4.setShowAsAction(
            MenuItem.SHOW_AS_ACTION_WITH_TEXT);
        mnu4.setCheckable(true);
    }
}

private boolean MenuChoice(MenuItem item)
{
    MainActivity mainActivity;
    mainActivity = (MainActivity) getActivity();

    switch (item.getItemId()) {
    case 0:
        Toast.makeText(getActivity(), "Tracking turned on",
            Toast.LENGTH_LONG).show();
        mainActivity.TrackingUsingGPS(true);
        mainActivity.TrackingUsingCellularWiFi(true);
        return true;
    case 1:
        Toast.makeText(getActivity(), "Tracking turned off",
            Toast.LENGTH_LONG).show();
        mainActivity.TrackingUsingGPS(false);
        mainActivity.TrackingUsingCellularWiFi(false);
        return true;
    case 2:
        if (!item.isChecked()) {
            Toast.makeText(getActivity(),
                "Using GPS for location tracking",
                Toast.LENGTH_LONG).show();
        }
        item.setChecked(!(item.isChecked()));
        mainActivity.TrackingUsingGPS(item.isChecked());
        return true;
    case 3:
        if (!item.isChecked()) {
            Toast.makeText(getActivity(),
                "Using Cellular/WiFi for location tracking",
                Toast.LENGTH_LONG).show();
        }
        item.setChecked(!(item.isChecked()));
        mainActivity.TrackingUsingCellularWiFi(item.isChecked());
        return true;
    }
    return false;
}
```

Note that when a particular option is selected (for example, Use GPS), a checkmark is displayed next to it (see Figure 4-20). This is achieved using the `setChecked()` method of the `MenuItem` object. In order for a checkmark to be displayed, you need to call its `setCheckable()` method first.

In order to call a method on the activity from within a fragment, you need to obtain an instance of the activity and then call its method directly, like this:

FIGURE 4-20

```
//---obtain an instance of the activity---
MainActivity mainActivity;
mainActivity = (MainActivity) getActivity();

//---calling the methods from the activity---
mainActivity.TrackingUsingGPS(true);
mainActivity.TrackingUsingCellularWiFi(true);
```

MONITORING A LOCATION

One very cool feature of the `LocationManager` class is its ability to monitor a specific location. This is achieved using the `addProximityAlert()` method. The following code snippet shows how to monitor a particular location so that if the user is within a five-meter radius from that location, your application will fire an intent to launch the web browser:

```
//--use the LocationManager class to obtain locations data--
lm = (LocationManager)
    getSystemService(Context.LOCATION_SERVICE);

//---PendingIntent to launch activity if the user is within
// some locations---
PendingIntent pendIntent = PendingIntent.getActivity(
    this, 0, new
    Intent(android.content.Intent.ACTION_VIEW,
      Uri.parse("http://www.amazon.com")), 0);

lm.addProximityAlert(37.422006, -122.084095, 5, -1, pendIntent);
```

The `addProximityAlert()` method takes five arguments: latitude, longitude, radius (in meters), expiration (time for the proximity alert to be valid, after which it will be deleted; -1 for no expiration), and the pending intent.

Note that if the Android device's screen goes to sleep, the proximity is also checked once every four minutes in order to preserve the battery life of the device.

SUMMARY

This chapter took a whirlwind tour of the `MapView` object, which displays Google Maps in your Android application. You have learned the various ways in which the map can be manipulated, and you have also learned how you can obtain geographical location data using the various network providers: GPS, Cellular triangulation, or Wi-Fi triangulation.

EXERCISES

1. If you have embedded the Google Maps API into your Android application but it does not show the map when the application is loaded, what could be the likely reasons?

2. What is the difference between geocoding and reverse geocoding?

3. Name the two location providers that you can use to obtain your location data.

4. What is the method for monitoring a location?

Answers to the Exercises can be found in Appendix C.

▶ WHAT YOU LEARNED IN THIS CHAPTER

TOPIC	KEY CONCEPTS

Displaying the MapView	`<com.google.android.maps.MapView`
	`    android:id="@+id/mapView"`
	`    android:layout_width="fill_parent"`
	`    android:layout_height="fill_parent"`
	`    android:enabled="true"`
	`    android:clickable="true"`
	`    android:apiKey="<your_key_here>" />`

Referencing the Map library	`<uses-library android:name="com.google.android.maps" />`
Displaying the zoom controls	`mapView.setBuiltInZoomControls(true);`
Programmatically zooming in or out of the map	`mc.zoomIn();` `mc.zoomOut();`
Changing views	`mapView.setSatellite(true);` `mapView.setTraffic(true);`

Animating to a particular location	`mc = mapView.getController();`
	`String coordinates[] = {"1.352566007",`
	`"103.78921587"};`
	`double lat = Double.`
	`parseDouble(coordinates[0]);`
	`double lng = Double.`
	`parseDouble(coordinates[1]);`
	`p = new GeoPoint(`
	`    (int) (lat * 1E6),`
	`    (int) (lng * 1E6));`
	` mc.animateTo(p);`

Adding markers	Implement an Overlay class and override the draw() method

Getting the location of the map touched	`GeoPoint p = mapView.getProjection().`
	`fromPixels(`
	`    (int) event.getX(),`
	`    (int) event.getY());`

Geocoding and reverse geocoding	Use the Geocoder class

TOPIC	KEY CONCEPTS
| **Obtaining location data** | ```java
private LocationManager lm;

//...

        lm = (LocationManager)
            getSystemService(Context.LOCATION_
SERVICE);

        locationListener = new
MyLocationListener();

        lm.requestLocationUpdates(
            LocationManager.GPS_PROVIDER,
            0,
            0,
            locationListener);

//...

    private class MyLocationListener implements
LocationListener
    {
        @Override
        public void onLocationChanged(Location
loc) {
            if (loc != null) {
            }
        }

        @Override
        public void onProviderDisabled(String
provider) {
        }

        @Override
        public void onProviderEnabled(String
provider) {
        }

        @Override
        public void onStatusChanged(String
provider, int status,
            Bundle extras) {
        }
    }
``` |
| **Monitoring a location** | ```java
lm.addProximityAlert(37.422006, -122.084095, 5,
-1, pendIntent);
``` |

SMS Messaging and Networking

WHAT YOU WILL LEARN IN THIS CHAPTER

➤ Sending SMS messages programmatically from within your application

➤ Sending SMS messages using the built-in Messaging application

➤ How to receive incoming SMS messages

➤ How to build a location tracker application

➤ Sending e-mail messages from your application

➤ Connecting to the Web using HTTP

➤ How to consume Web services

Once your basic Android application is up and running, the next interesting thing you can add to it is the capability to communicate with the outside world. You may want your application to send an SMS message to another phone when an event happens (such as when you reach a particular geographical location), or you may wish to access a Web service that provides certain services (such as currency exchange, weather, etc.). In this chapter, you learn how to send and receive SMS messages programmatically from within your Android application.

You will also learn how to use the HTTP protocol to talk to web servers so that you can download text and binary data. The last part of this chapter shows you how to parse XML files to extract the relevant parts of an XML file — a technique that is useful if you are accessing Web services.

SMS MESSAGING

SMS messaging is one of the main *killer applications* on a mobile device today. Any mobile phone (or devices) you buy today should have at least SMS messaging capabilities, and nearly all users of any age know how to send and receive such messages. Android comes with a built-in SMS application that enables you to send and receive SMS messages. However, in some cases you might want to integrate SMS capabilities into your own Android application. For example, you might want to write an application that automatically sends an SMS message at regular time intervals. This would be useful, for example, if you wanted to track the location of your kids — simply give them an Android device that sends out an SMS message containing its geographical location every 30 minutes. Now you know if they really went to the library after school! (Of course, that would also mean you would have to pay the fees incurred for sending all those SMS messages . . .)

This section describes how you can programmatically send and receive SMS messages in your Android applications. The good news for Android developers is that you don't need a real device to test SMS messaging: The free Android emulator provides that capability.

Sending SMS Messages Programmatically

You will first learn how to send SMS messages programmatically from within your application. Using this approach, your application can automatically send an SMS message to a recipient without user intervention. The following Try It Out shows you how.

> **NOTE** *For the following Try It Out, you will create an Android 2.2 application because it is easier and faster to test it out on a pre-Android 3.0 emulator. The concepts covered apply to Android 3.0 as well. Later in this chapter, you will modify the project to run on Android 3.0 devices.*

TRY IT OUT Sending SMS Messages

codefile SMS.zip available for download at Wrox.com

1. Using Eclipse, create a new Android project and name it **SMS**, as shown in Figure 5-1.

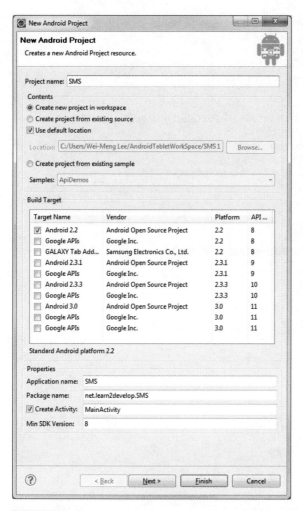

FIGURE 5-1

2. Add the following statements in bold to the `main.xml` file:

```xml
<?xml version="1.0" encoding="utf-8"?>
<LinearLayout xmlns:android="http://schemas.android.com/apk/res/android"
    android:orientation="vertical"
    android:layout_width="fill_parent"
    android:layout_height="fill_parent"
>
<Button
    android:id="@+id/btnSendSMS"
    android:layout_width="fill_parent"
    android:layout_height="wrap_content"
    android:text="Send SMS" />
</LinearLayout>
```

3.   In the `AndroidManifest.xml` file, add the following statements in bold:

```xml
<?xml version="1.0" encoding="utf-8"?>
<manifest xmlns:android="http://schemas.android.com/apk/res/android"
      package="net.learn2develop.SMS"
      android:versionCode="1"
      android:versionName="1.0">
    <application android:icon="@drawable/icon" android:label="@string/app_name">
        <activity android:name=".MainActivity"
                  android:label="@string/app_name">
            <intent-filter>
                <action android:name="android.intent.action.MAIN" />
                <category android:name="android.intent.category.LAUNCHER" />
            </intent-filter>
        </activity>
    </application>
    <uses-sdk android:minSdkVersion="8" />
    <uses-permission android:name="android.permission.SEND_SMS">
    </uses-permission>
</manifest>
```

4.   Add the following statements in bold to the `MainActivity.java` file:

```java
package net.learn2develop.SMS;

import android.app.Activity;
import android.os.Bundle;

import android.app.PendingIntent;
import android.content.Intent;
import android.telephony.SmsManager;
import android.view.View;
import android.widget.Button;

public class MainActivity extends Activity {
    Button btnSendSMS;
    /** Called when the activity is first created. */
    @Override
    public void onCreate(Bundle savedInstanceState) {
        super.onCreate(savedInstanceState);
        setContentView(R.layout.main);

        btnSendSMS = (Button) findViewById(R.id.btnSendSMS);
        btnSendSMS.setOnClickListener(new View.OnClickListener()
        {
            public void onClick(View v)
            {
                sendSMS("5556", "Hello my friends!");

            }
        });
    }

    //---sends an SMS message to another device---
```

```
private void sendSMS(String phoneNumber, String message)
{
    SmsManager sms = SmsManager.getDefault();
    sms.sendTextMessage(phoneNumber, null, message, null, null);
}
}
```

5. Press F11 to debug the application on an Android 2.2 emulator. Using the Android SDK and AVD Manager, launch another Android 2.2 emulator.

6. On the first Android emulator, click the Send SMS button to send an SMS message to the second emulator. The left side of Figure 5-2 shows the SMS message received by the second emulator (note the notification bar at the top of the second emulator).

FIGURE 5-2

How It Works

Android uses a permissions-based policy whereby all the permissions needed by an application must be specified in the AndroidManifest.xml file. This ensures that when the application is installed, the user knows exactly which access permissions it requires.

Because sending SMS messages incurs additional costs on the user's end, indicating the SMS permissions in the `AndroidManifest.xml` file enables users to decide whether to allow the application to install or not.

To send an SMS message programmatically, you used the `SmsManager` class. Unlike other classes, you do not directly instantiate this class; instead, you call the `getDefault()` static method to obtain an `SmsManager` object. You then sent the SMS message using the `sendTextMessage()` method:

```
private void sendSMS(String phoneNumber, String message)
{
    SmsManager sms = SmsManager.getDefault();
    sms.sendTextMessage(phoneNumber, null, message, null, null);
}
```

Following are the five arguments to the `sendTextMessage()` method:

➤ `destinationAddress` — Phone number of the recipient.

➤ `scAddress` — Service center address; use null for default SMSC.

➤ `text` — Content of the SMS message.

➤ `sentIntent` — Pending intent to invoke when the message is sent (discussed in more detail in the next section).

➤ `deliveryIntent` — Pending intent to invoke when the message has been delivered (discussed in more detail in the next section).

Getting Feedback after Sending the Message

In the previous section, you learned how to programmatically send SMS messages using the `SmsManager` class; but how do you know that the message has been sent correctly? To do so, you can create two `PendingIntent` objects to monitor the status of the SMS message-sending process. These two `PendingIntent` objects are passed to the last two arguments of the `sendTextMessage()` method. The following code snippets show how you can monitor the status of the SMS message being sent:

```
//---sends an SMS message to another device---
private void sendSMS(String phoneNumber, String message)
{
    String SENT = "SMS_SENT";
    String DELIVERED = "SMS_DELIVERED";

    PendingIntent sentPI = PendingIntent.getBroadcast(this, 0,
        new Intent(SENT), 0);

    PendingIntent deliveredPI = PendingIntent.getBroadcast(this, 0,
        new Intent(DELIVERED), 0);

    //---when the SMS has been sent---
```

```java
        registerReceiver(new BroadcastReceiver(){
            @Override
            public void onReceive(Context arg0, Intent arg1) {
                switch (getResultCode())
                {
                    case Activity.RESULT_OK:
                        Toast.makeText(getBaseContext(), "SMS sent",
                                Toast.LENGTH_SHORT).show();
                        break;
                    case SmsManager.RESULT_ERROR_GENERIC_FAILURE:
                        Toast.makeText(getBaseContext(), "Generic failure",
                                Toast.LENGTH_SHORT).show();
                        break;
                    case SmsManager.RESULT_ERROR_NO_SERVICE:
                        Toast.makeText(getBaseContext(), "No service",
                                Toast.LENGTH_SHORT).show();
                        break;
                    case SmsManager.RESULT_ERROR_NULL_PDU:
                        Toast.makeText(getBaseContext(), "Null PDU",
                                Toast.LENGTH_SHORT).show();
                        break;
                    case SmsManager.RESULT_ERROR_RADIO_OFF:
                        Toast.makeText(getBaseContext(), "Radio off",
                                Toast.LENGTH_SHORT).show();
                        break;
                }
            }
        }, new IntentFilter(SENT));

        //---when the SMS has been delivered---
        registerReceiver(new BroadcastReceiver(){
            @Override
            public void onReceive(Context arg0, Intent arg1) {
                switch (getResultCode())
                {
                    case Activity.RESULT_OK:
                        Toast.makeText(getBaseContext(), "SMS delivered",
                                Toast.LENGTH_SHORT).show();
                        break;
                    case Activity.RESULT_CANCELED:
                        Toast.makeText(getBaseContext(), "SMS not delivered",
                                Toast.LENGTH_SHORT).show();
                        break;
                }
            }
        }, new IntentFilter(DELIVERED));

        SmsManager sms = SmsManager.getDefault();
        sms.sendTextMessage(phoneNumber, null, message, sentPI, deliveredPI);
}
```

Here, you created two `PendingIntent` objects. You then registered for two `BroadcastReceivers`. These two `BroadcastReceivers` listen for intents that match "SMS_SENT" and "SMS_DELIVERED" (which are fired by the OS when the message has been sent and delivered, respectively). Within each `BroadcastReceiver` you override the `onReceive()` method and get the current result code.

The two `PendingIntent` objects are passed into the last two arguments of the `sendTextMessage()` method:

```
sms.sendTextMessage(phoneNumber, null, message, sentPI, deliveredPI);
```

In this case, whether a message has been sent correctly or failed to be delivered, you will be notified of its status via the two `PendingIntent` objects.

Sending SMS Messages Using Intent

Using the `SmsManager` class, you can send SMS messages from within your application without the need to involve the built-in Messaging application. However, sometimes it would be easier if you could simply invoke the built-in Messaging application and let it do all the work of sending the message.

To activate the built-in Messaging application from within your application, you can use an `Intent` object together with the MIME type `"vnd.android-dir/mms-sms"`, as shown by the following code snippet:

```
/** Called when the activity is first created. */
@Override
public void onCreate(Bundle savedInstanceState) {
    super.onCreate(savedInstanceState);
    setContentView(R.layout.main);

    btnSendSMS = (Button) findViewById(R.id.btnSendSMS);
    btnSendSMS.setOnClickListener(new View.OnClickListener()
    {
        public void onClick(View v)
        {
            //sendSMS("5556", "Hello my friends!");
            Intent i = new
                Intent(android.content.Intent.ACTION_VIEW);
            i.putExtra("address", "5556; 5558; 5560");
            i.putExtra("sms_body", "Hello my friends!");
            i.setType("vnd.android-dir/mms-sms");
            startActivity(i);
        }
    });
}
```

This will invoke the Messaging application, as shown in Figure 5-3. Note that you can send your SMS to multiple recipients by simply separating each phone number with a semicolon (in the `putExtra()` method).

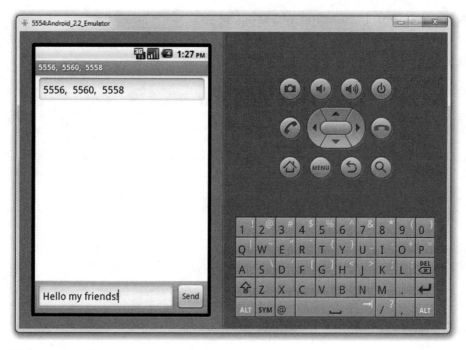

FIGURE 5-3

NOTE *If you use this method to invoke the Messaging application, there is no need to ask for the* SMS_SEND *permission in* AndroidManifest.xml *because your application is ultimately not the one sending the message.*

Receiving SMS Messages

Besides sending SMS messages from your Android applications, you can also receive incoming SMS messages from within your application by using a BroadcastReceiver object. This is useful when you want your application to perform an action when a certain SMS message is received. For example, you might want to track the location of your phone in case it is lost or stolen. In this case, you can write an application that automatically listens for SMS messages containing some secret code. Once that message is received, you can then send an SMS message containing the location's coordinates back to the sender.

The following Try It Out shows how to programmatically listen for incoming SMS messages.

TRY IT OUT Intercepting Incoming SMS Messages

1. Using the same project created in the previous section, add the following statements in bold to the `AndroidManifest.xml` file:

```xml
<?xml version="1.0" encoding="utf-8"?>
<manifest xmlns:android="http://schemas.android.com/apk/res/android"
      package="net.learn2develop.SMS"
      android:versionCode="1"
      android:versionName="1.0">
   <application android:icon="@drawable/icon" android:label="@string/app_name">
      <activity android:name=".MainActivity"
                 android:label="@string/app_name">
         <intent-filter>
            <action android:name="android.intent.action.MAIN" />
            <category android:name="android.intent.category.LAUNCHER" />
         </intent-filter>
      </activity>
      <receiver android:name=".SMSReceiver">
         <intent-filter>
            <action android:name=
                "android.provider.Telephony.SMS_RECEIVED" />
         </intent-filter>
      </receiver>
   </application>
   <uses-sdk android:minSdkVersion="8" />
   <uses-permission android:name="android.permission.SEND_SMS"></uses-permission>
   <uses-permission android:name="android.permission.RECEIVE_SMS">
   </uses-permission>
</manifest>
```

2. In the `src` folder of the project, add a new class file to the package name and call it `SMSReceiver.java` (see Figure 5-4).

3. Code the `SMSReceiver.java` file as follows:

FIGURE 5-4

```java
package net.learn2develop.SMS;

import android.content.BroadcastReceiver;
import android.content.Context;
import android.content.Intent;
import android.os.Bundle;
import android.telephony.SmsMessage;
import android.widget.Toast;

public class SMSReceiver extends BroadcastReceiver
{
    @Override
    public void onReceive(Context context, Intent intent)
    {
        //---get the SMS message passed in---
        Bundle bundle = intent.getExtras();
        SmsMessage[] msgs = null;
        String str = "";
```

```
        if (bundle != null)
        {
            //---retrieve the SMS message received---
            Object[] pdus = (Object[]) bundle.get("pdus");
            msgs = new SmsMessage[pdus.length];
            for (int i=0; i<msgs.length; i++){
                msgs[i] = SmsMessage.createFromPdu((byte[])pdus[i]);
                str += "SMS from " + msgs[i].getOriginatingAddress();
                str += " :";
                str += msgs[i].getMessageBody().toString();
                str += "\n";
            }
            //---display the new SMS message---
            Toast.makeText(context, str, Toast.LENGTH_SHORT).show();
        }
    }
}
```

4. Press F11 to debug the application on the Android emulator.

5. Using the DDMS, send a message to the emulator. Your application should be able to receive the message and display it using the Toast class (see Figure 5-5).

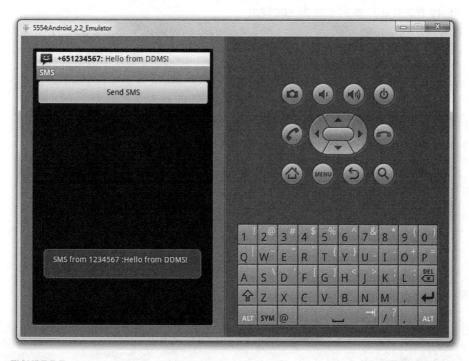

FIGURE 5-5

How It Works

To listen for incoming SMS messages, you created a `BroadcastReceiver` class. The `BroadcastReceiver` class enabled your application to receive intents sent by other applications using the `sendBroadcast()` method. Essentially, the `sendBroadcast()` method enables your application to handle events raised by other applications. When an intent is received, the `onReceive()` method is called; hence, you need to override this.

When an incoming SMS message is received, the `onReceive()` method is fired. The SMS message is contained in the `Intent` object (`intent`; the second parameter in the `onReceive()` method) via a `Bundle` object. The messages are stored in an `Object` array in the PDU format. To extract each message, you use the static `createFromPdu()` method from the `SmsMessage` class. The SMS message is then displayed using the `Toast` class. The phone number of the sender is obtained via the `getOriginatingAddress()` method, so if you need to send an autoreply to the sender, this is the method to obtain the sender's phone number.

One interesting characteristic of the `BroadcastReceiver` is that you can continue to listen for incoming SMS messages even if the application is not running; as long as the application is installed on the device, any incoming SMS messages will be received by the application.

Updating an Activity from a BroadcastReceiver

The previous section described how you can use a `BroadcastReceiver` class to listen for incoming SMS messages and then use the `Toast` class to display the received SMS message. Often, you'll want to send the SMS message back to the main activity of your application. For example, you might wish to display the message in a `TextView`. The following Try It Out demonstrates how you can do this.

TRY IT OUT Updating an Activity through a BroadcastReceiver

1. Using the same project created in the previous section, add the following lines in bold to the `main.xml` file:

```xml
<?xml version="1.0" encoding="utf-8"?>
<LinearLayout xmlns:android="http://schemas.android.com/apk/res/android"
    android:orientation="vertical"
    android:layout_width="fill_parent"
    android:layout_height="fill_parent"
>
<Button
    android:id="@+id/btnSendSMS"
    android:layout_width="fill_parent"
    android:layout_height="wrap_content"
    android:text="Send SMS" />

<TextView
    android:id="@+id/textView1"
    android:layout_width="wrap_content"
    android:layout_height="wrap_content" />

</LinearLayout>
```

2. Add the following statements in bold to the SMSReceiver.java file:

```java
package net.learn2develop.SMS;

import android.content.BroadcastReceiver;
import android.content.Context;
import android.content.Intent;
import android.os.Bundle;
import android.telephony.SmsMessage;
import android.widget.Toast;

public class SMSReceiver extends BroadcastReceiver
{
    @Override
    public void onReceive(Context context, Intent intent)
    {
        //---get the SMS message passed in---
        Bundle bundle = intent.getExtras();
        SmsMessage[] msgs = null;
        String str = "";
        if (bundle != null)
        {
            //---retrieve the SMS message received---
            Object[] pdus = (Object[]) bundle.get("pdus");
            msgs = new SmsMessage[pdus.length];
            for (int i=0; i<msgs.length; i++){
                msgs[i] = SmsMessage.createFromPdu((byte[])pdus[i]);
                str += "SMS from " + msgs[i].getOriginatingAddress();
                str += " :";
                str += msgs[i].getMessageBody().toString();
                str += "\n";
            }
            //---display the new SMS message---
            Toast.makeText(context, str, Toast.LENGTH_SHORT).show();

            //---send a broadcast intent to update the SMS received in
            // the activity---
            Intent broadcastIntent = new Intent();
            broadcastIntent.setAction("SMS_RECEIVED_ACTION");
            broadcastIntent.putExtra("sms", str);
            context.sendBroadcast(broadcastIntent);
        }
    }
}
```

3. Add the following statements in bold to the MainActivity.java file:

```java
package net.learn2develop.SMS;

import android.app.Activity;
import android.os.Bundle;
import android.app.PendingIntent;
import android.content.Context;
import android.content.Intent;
```

```java
import android.telephony.SmsManager;
import android.view.View;
import android.widget.Button;
import android.widget.Toast;


import android.content.BroadcastReceiver;
import android.content.IntentFilter;
import android.widget.TextView;

public class MainActivity extends Activity {
    Button btnSendSMS;
    IntentFilter intentFilter;

    private BroadcastReceiver intentReceiver = new BroadcastReceiver() {
        @Override
        public void onReceive(Context context, Intent intent) {
            //---display the SMS received in the TextView---
            TextView SMSes = (TextView) findViewById(R.id.textView1);
            SMSes.setText(intent.getExtras().getString("sms"));
        }
    };

    /** Called when the activity is first created. */
    @Override
    public void onCreate(Bundle savedInstanceState) {
        super.onCreate(savedInstanceState);
        setContentView(R.layout.main);

        //---intent to filter for SMS messages received---
        intentFilter = new IntentFilter();
        intentFilter.addAction("SMS_RECEIVED_ACTION");

        btnSendSMS = (Button) findViewById(R.id.btnSendSMS);
        btnSendSMS.setOnClickListener(new View.OnClickListener()
        {
            public void onClick(View v)
            {
                //sendSMS("5554", "Hello my friends!");

                Intent i = new
                    Intent(android.content.Intent.ACTION_VIEW);
                i.putExtra("address", "5556; 5558; 5560");
                i.putExtra("sms_body", "Hello my friends!");
                i.setType("vnd.android-dir/mms-sms");
                startActivity(i);
            }
        });
    }

    @Override
    protected void onResume() {
        //---register the receiver---
        registerReceiver(intentReceiver, intentFilter);
```

```
        super.onResume();
    }

    @Override
    protected void onPause() {
        //---unregister the receiver---
        unregisterReceiver(intentReceiver);
        super.onPause();
    }

    //---sends an SMS message to another device---
    private void sendSMS(String phoneNumber, String message)
    {
        //...
    }
}
```

4. Press F11 to debug the application on the Android emulator. Using the DDMS, send an SMS message to the emulator. Figure 5-6 shows the Toast class displaying the message received, and the TextView showing the message received.

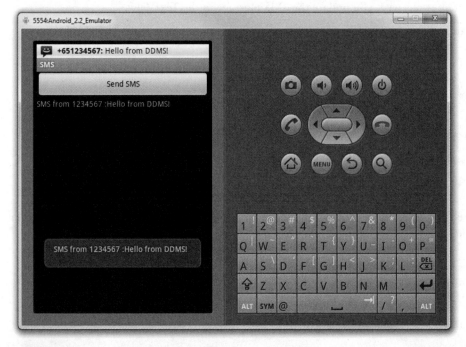

FIGURE 5-6

How It Works

You first added a TextView to your activity so that it can be used to display the received SMS message.

Next, you modified the SMSReceiver class so that when it receives an SMS message, it will broadcast another Intent object so that any applications listening for this intent can be notified (which we will implement in the activity next). The SMS received is also sent out via this intent:

```
//---send a broadcast intent to update the SMS received in
// the activity---
Intent broadcastIntent = new Intent();
broadcastIntent.setAction("SMS_RECEIVED_ACTION");
broadcastIntent.putExtra("sms", str);
context.sendBroadcast(broadcastIntent);
```

In your activity, you then created a BroadcastReceiver object to listen for broadcast intents:

```
private BroadcastReceiver intentReceiver = new BroadcastReceiver() {
    @Override
    public void onReceive(Context context, Intent intent) {
        //---display the SMS received in the TextView---
        TextView SMSes = (TextView) findViewById(R.id.textView1);
        SMSes.setText(intent.getExtras().getString("sms"));
    }
};
```

When a broadcast intent is received, you update the SMS message in the TextView.

You need to create an IntentFilter object so that you can listen for a particular intent. In this case, the intent is "SMS_RECEIVED_ACTION":

```
@Override
public void onCreate(Bundle savedInstanceState) {
    super.onCreate(savedInstanceState);
    setContentView(R.layout.main);

    //---intent to filter for SMS messages received---
    intentFilter = new IntentFilter();
    intentFilter.addAction("SMS_RECEIVED_ACTION");
    //...
}
```

Finally, you registered the BroadcastReceiver in the activity's onResume() event and unregistered it in the onPause() event:

```
@Override
protected void onResume() {
    //---register the receiver---
    registerReceiver(intentReceiver, intentFilter);
    super.onResume();
}

@Override
protected void onPause() {
    //---unregister the receiver---
    unregisterReceiver(intentReceiver);
    super.onPause();
}
```

This means that the TextView will display the SMS message only when the message is received while the activity is visible on the screen. If the SMS message is received when the activity is not in the foreground, the TextView will not be updated.

Invoking an Activity from a BroadcastReceiver

The previous example showed how you can pass the SMS message received to be displayed in the activity. However, in many situations your activity may be in the background when the SMS message is received. In this case, it would be useful to be able to bring the activity to the foreground when a message is received. The following Try It Out shows you how.

TRY IT OUT   **Invoking an Activity**

1. Using the same project created earlier, add the following lines in bold to the `MainActivity` `.java` file:

```java
/** Called when the activity is first created. */
@Override
public void onCreate(Bundle savedInstanceState) {
    super.onCreate(savedInstanceState);
    setContentView(R.layout.main);

    //---intent to filter for SMS messages received---
    intentFilter = new IntentFilter();
    intentFilter.addAction("SMS_RECEIVED_ACTION");

    //---register the receiver---
    registerReceiver(intentReceiver, intentFilter);

    btnSendSMS = (Button) findViewById(R.id.btnSendSMS);
    btnSendSMS.setOnClickListener(new View.OnClickListener()
    {
        public void onClick(View v)
        {
            //sendSMS("5554", "Hello my friends!");
            Intent i = new
                Intent(android.content.Intent.ACTION_VIEW);
            i.putExtra("address", "5556; 5558; 5560");
            i.putExtra("sms_body", "Hello my friends!");
            i.setType("vnd.android-dir/mms-sms");
            startActivity(i);
        }
    });
}

@Override
protected void onResume() {
    //---register the receiver---
    //registerReceiver(intentReceiver, intentFilter);
```

```
        super.onResume();
    }

    @Override
    protected void onPause() {
        //---unregister the receiver---
        //unregisterReceiver(intentReceiver);
        super.onPause();
    }

    @Override
    protected void onDestroy() {
        //---unregister the receiver---
        unregisterReceiver(intentReceiver);
        super.onPause();
    }
```

2. Add the following statements in bold to the SMSReceiver.java file:

```
    @Override
    public void onReceive(Context context, Intent intent)
    {
        //---get the SMS message passed in---
        Bundle bundle = intent.getExtras();
        SmsMessage[] msgs = null;
        String str = "";
        if (bundle != null)
        {
            //---retrieve the SMS message received---
            Object[] pdus = (Object[]) bundle.get("pdus");
            msgs = new SmsMessage[pdus.length];
            for (int i=0; i<msgs.length; i++){
                msgs[i] = SmsMessage.createFromPdu((byte[])pdus[i]);
                str += "SMS from " + msgs[i].getOriginatingAddress();
                str += " :";
                str += msgs[i].getMessageBody().toString();
                str += "\n";
            }
            //---display the new SMS message---
            Toast.makeText(context, str, Toast.LENGTH_SHORT).show();

            //---launch the MainActivity---
            Intent mainActivityIntent = new Intent(context, MainActivity.class);
            mainActivityIntent.setFlags(Intent.FLAG_ACTIVITY_NEW_TASK);
            context.startActivity(mainActivityIntent);

            //---send a broadcast to update the SMS received in the activity---
            Intent broadcastIntent = new Intent();
            broadcastIntent.setAction("SMS_RECEIVED_ACTION");
            broadcastIntent.putExtra("sms", str);
            context.sendBroadcast(broadcastIntent);
        }
    }
```

3. Modify the `main.xml` file as follows:

```
<activity android:name=".MainActivity"
        android:label="@string/app_name"
        android:launchMode="singleTask" >
    <intent-filter>
        <action android:name="android.intent.action.MAIN" />
        <category android:name="android.intent.category.LAUNCHER" />
    </intent-filter>
</activity>
```

4. Press F11 to debug the application on the Android emulator. When the `MainActivity` is shown, click the Home button to send the activity to the background.

5. Use the DDMS to send an SMS message to the emulator again. This time, note that the activity is brought to the foreground, displaying the SMS message received.

How It Works

In the `MainActivity` class, you first registered the `BroadcastReceiver` in the activity's `onCreate()` event, instead of the `onResume()` event; and instead of unregistering it in the `onPause()` event, you unregistered it in the `onDestroy()` event. This ensures that even if the activity is in the background, it can still listen for the broadcast intent.

Next, you modified the `onReceive()` event in the `SMSReceiver` class by using an intent to bring the activity to the foreground before broadcasting another intent:

```
//---launch the MainActivity---
Intent mainActivityIntent =
    new Intent(context, MainActivity.class);
mainActivityIntent.setFlags(Intent.FLAG_ACTIVITY_NEW_TASK);
context.startActivity(mainActivityIntent);

//---send a broadcast to update the SMS received in the activity---
Intent broadcastIntent = new Intent();
broadcastIntent.setAction("SMS_RECEIVED_ACTION");
broadcastIntent.putExtra("sms", str);
context.sendBroadcast(broadcastIntent);
```

The `startActivity()` method launches the activity and brings it to the foreground. Note that you needed to set the `Intent.FLAG_ACTIVITY_NEW_TASK` flag because calling `startActivity()` from outside of an activity context requires the `FLAG_ACTIVITY_NEW_TASK` flag.

You also needed to set the `launchMode` attribute of the `<activity>` element in the `AndroidManifest.xml` file to `singleTask`:

```
<activity android:name=".MainActivity"
        android:label="@string/app_name"
        android:launchMode="singleTask" >
```

If you don't set this, multiple instances of the activity will be launched as your application receives SMS messages.

Note that in this example, when the activity is in the background (such as when you click the Home button to show the home screen), the activity was brought to the foreground and its TextView was updated with the SMS received. However, if the activity were killed (such as when you click the Back button to destroy it), the activity is launched again but the TextView is not updated.

Caveats and Warnings

While the capability to send and receive SMS messages makes Android a very compelling platform for developing sophisticated applications, this flexibility comes with a price. A seemingly innocent application may send SMS messages behind the scenes without the user knowing, as demonstrated by a recent case of an SMS-based Trojan Android application (`http://forum.vodafone.co.nz/topic/5719-android-sms-trojan-warning/`). Claiming to be a media player, after it is installed the application sends SMS messages to a premium number, resulting in huge phone bills for the user.

While the user needs to explicitly give permission to your application, the request for permission is only shown at installation time. Figure 5-7 shows the request for permission that appears when you try to install the application (as an APK file; Chapter6 discusses packaging your Android applications in more detail) on the emulator (same as on a real device). If the user clicks the Install button, he or she is considered to have given permission to allow the application to send and receive SMS messages. This is dangerous, because after the application is installed it can send and receive SMS messages without ever prompting the user again.

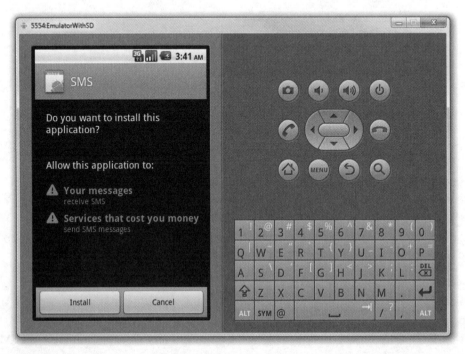

FIGURE 5-7

In addition to this, the application can also "sniff" for incoming SMS messages. For example, based on the techniques you learned from the previous section, you can easily write an application that checks for certain keywords in the SMS message. When an SMS message contains the keyword you are looking for, you can use the Location Manager (covered in Chapter 4) to obtain your geographical location and then send the coordinates back to the sender of the SMS message. The sender could then easily track your location. All these tasks can be done without the user knowing it! That said, users should try to avoid installing Android applications that come from dubious sources, such as unknown websites, strangers, and so on.

Project: Building the Location Tracker Applications

Now that you have learned how to make your Android application send and receive SMS messages, let's put this knowledge to good use by writing a pair of applications that enable users to track the whereabouts of their friends. We'll call the pair of applications — one for the phone and one for the tablet — the *Location Tracker*.

The phone application will listen for incoming SMS messages. If an incoming message starts with the sentence "`Where are you?`", it will invoke the Location Manager (see Chapter 4) to obtain the device's geographical location. It will then send back an SMS message (to the sender) containing the device's latitude and longitude. The format of the message is "`location:<lat>:<lon>.`"

The tablet application, conversely, allows users to send SMS messages to the phone application and wait for returning SMS messages containing the devices' locations. It, too, waits for incoming messages, but only for those messages that start with the word "`location:.`" Once these messages are received, it will display the location of the device using Google Maps. Figure 5-8 summarizes the flow of the two applications.

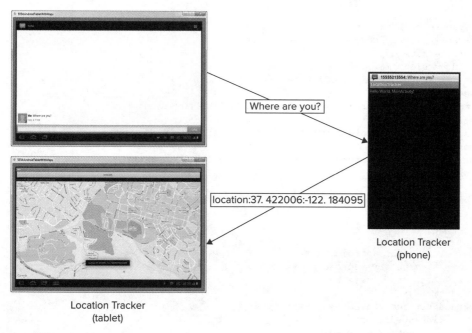

Where are you?

location:37. 422006:-122. 184095

Location Tracker
(phone)

Location Tracker
(tablet)

FIGURE 5-8

WARNING *In some countries, it is illegal for you to track the location of a person without his or her knowledge. If you install the Location Tracker application on a user's phone, that device will automatically return its location information to whomever sends it an SMS message beginning with the words "Where are you?" Therefore, if you want to use this project in real life, you must alert potential users about the application's functionality, so that they have the option to not reveal their location. That said, this project also has its perfectly legitimate uses – using it to track your child while they're on a field trip, or using it to track your elderly folks when they go travelling, for example.*

Building the Location Tracker (Phone)

The first application you will build for this project is the application that will be installed on an Android phone. Once it is installed, you will be able to track its whereabouts by sending it an SMS message. The following Try It Out shows you how.

TRY IT OUT Creating the Location Tracker Application (Phone)

1. Using Eclipse, create a new Android application and name it `LocationTracker`. For this project, select the Android 2.2 target, as the Android 2.3 emulator seems to have problems receiving SMS messages from the DDMS perspective.

2. Modify the `AndroidManifest.xml` as follows:

```xml
<?xml version="1.0" encoding="utf-8"?>
<manifest
    xmlns:android="http://schemas.android.com/apk/res/android"
    package="net.learn2develop.LocationTracker"
    android:versionCode="1"
    android:versionName="1.0">
    <uses-sdk android:minSdkVersion="8" />
    <uses-permission
        android:name="android.permission.ACCESS_FINE_LOCATION">
    </uses-permission>
    <uses-permission android:name="android.permission.RECEIVE_SMS">
    </uses-permission>
    <uses-permission android:name="android.permission.SEND_SMS">
    </uses-permission>
<application
    android:icon="@drawable/icon" android:label="@string/app_name">
        <activity android:name=".MainActivity"
                android:label="@string/app_name">
            <intent-filter>
                <action android:name="android.intent.action.MAIN" />
                <category android:name="android.intent.category.LAUNCHER" />
            </intent-filter>
        </activity>
        <receiver android:name=".SMSReceiver">
            <intent-filter>
                <action android:name=
```

```
                    "android.provider.Telephony.SMS_RECEIVED" />
            </intent-filter>
        </receiver>
    </application>
</manifest>
```

3. Add a class file under the package name and name it SMSReceiver.java. Populate it as follows:

```java
package net.learn2develop.LocationTracker;

import android.content.BroadcastReceiver;
import android.content.Context;
import android.content.Intent;
import android.os.Bundle;
import android.telephony.SmsMessage;

public class SMSReceiver extends BroadcastReceiver
{
    @Override
    public void onReceive(Context context, Intent intent)
    {
        //---get the SMS message that was received---
        Bundle bundle = intent.getExtras();
        SmsMessage[] msgs = null;
        String str = "";
        if (bundle != null)
        {
            //---retrieve the SMS message received---
            Object[] pdus = (Object[]) bundle.get("pdus");
            msgs = new SmsMessage[pdus.length];
            String senderTel = "";
            for (int i=0; i<msgs.length; i++){
                msgs[i] = SmsMessage.createFromPdu((byte[])pdus[i]);
                //---store the sender phone number---
                senderTel = msgs[i].getOriginatingAddress();
                //---get the body of the message received---
                str += msgs[i].getMessageBody().toString();
            }

            if (str.startsWith("Where are you?")) {
                //---send a broadcast to update the SMS received in
                // the activity---
                Intent broadcastIntent = new Intent();
                broadcastIntent.putExtra("senderTel", senderTel);
                broadcastIntent.setAction("SMS_RECEIVED_ACTION");
                context.sendBroadcast(broadcastIntent);
            }
        }
    }
}
```

4. Modify MainActivity.java as follows:

```java
package net.learn2develop.LocationTracker;

import android.app.Activity;
```

```java
import android.os.Bundle;

import android.content.BroadcastReceiver;
import android.content.Context;
import android.content.Intent;
import android.content.IntentFilter;
import android.location.Location;
import android.location.LocationListener;
import android.location.LocationManager;
import android.telephony.SmsManager;

public class MainActivity extends Activity {
    private IntentFilter intentFilter;
    private LocationManager lm;
    private LocationListener locationListener;
    private String senderTel;

    /** Called when the activity is first created. */
    @Override
    public void onCreate(Bundle savedInstanceState) {
        super.onCreate(savedInstanceState);
        setContentView(R.layout.main);

        //---intent to filter for SMS messages received---
        intentFilter = new IntentFilter();
        intentFilter.addAction("SMS_RECEIVED_ACTION");

        //---register the receiver---
        registerReceiver(intentReceiver, intentFilter);
    }

    private BroadcastReceiver intentReceiver = new BroadcastReceiver() {
        @Override
        public void onReceive(Context context, Intent intent) {
            //---get the phone number of the sender passed in via
            // the intent---
            senderTel = intent.getExtras().getString("senderTel");

            //---use the LocationManager class to obtain locations data---
            lm = (LocationManager)
                getSystemService(Context.LOCATION_SERVICE);

            //---request location updates---
            locationListener = new MyLocationListener();
            lm.requestLocationUpdates(
                LocationManager.GPS_PROVIDER,
                0,
                0,
                locationListener);
        }
    };

    private class MyLocationListener implements LocationListener
    {
        @Override
```

```
public void onLocationChanged(Location loc) {
    if (loc != null) {
        //---send a SMS containing the current location---
        SmsManager sms = SmsManager.getDefault();
        sms.sendTextMessage(senderTel, null,
            "location:" + loc.getLatitude() + ":" +
            loc.getLongitude(), null, null);
        //---stop listening for location changes---
        lm.removeUpdates(locationListener);
    }
}

@Override
public void onProviderDisabled(String provider) {
}

@Override
public void onProviderEnabled(String provider) {
}

@Override
public void onStatusChanged(String provider, int status,
    Bundle extras) {
}
    }
}
```

That's it for now. You will learn how to test the application later, after the tablet application is built.

How It Works

Basically, the application you built here simply listens for incoming SMS messages. The SMSReceiver class listens for incoming SMS messages. If the message starts with the sentence "Where are you?", it sends a broadcast (SMS_RECEIVED_ACTION) containing the sender's phone number:

```
//---retrieve the SMS message received---
Object[] pdus = (Object[]) bundle.get("pdus");
msgs = new SmsMessage[pdus.length];
String senderTel = "";
for (int i=0; i<msgs.length; i++){
    msgs[i] = SmsMessage.createFromPdu((byte[])pdus[i]);
    //---store the sender phone number---
    senderTel = msgs[i].getOriginatingAddress();
    //---get the body of the message received---
    str += msgs[i].getMessageBody().toString();
}

if (str.startsWith("Where are you?")) {
    //---send a broadcast to update the SMS received in
    // the activity---
    Intent broadcastIntent = new Intent();
    broadcastIntent.putExtra("senderTel", senderTel);
    broadcastIntent.setAction("SMS_RECEIVED_ACTION");
    context.sendBroadcast(broadcastIntent);
}
```

The `MainActivity` listens for this `SMS_RECEIVED_ACTION` intent:

```
private BroadcastReceiver intentReceiver = new BroadcastReceiver() {
    @Override
    public void onReceive(Context context, Intent intent) {
        //---get the phone number of the sender passed in via
        // the intent---
        senderTel = intent.getExtras().getString("senderTel");

        //---use the LocationManager class to obtain locations data---
        lm = (LocationManager)
            getSystemService(Context.LOCATION_SERVICE);

        //---request location updates---
        locationListener = new MyLocationListener();
        lm.requestLocationUpdates(
            LocationManager.GPS_PROVIDER,
            0,
            0,
            locationListener);
    }
};
```

Once this intent is received, it extracts the sender's phone number so that later you can send an SMS message containing the user's location back to the sender. Here, you requested for location updates using the GPS provider (alternatively, you can also use the `NETWORK_PROVIDER` as shown in Chapter 4). When a location is obtained, you send the location information using SMS:

```
@Override
public void onLocationChanged(Location loc) {
    if (loc != null) {
        //---send a SMS containing the current location---
        SmsManager sms = SmsManager.getDefault();
        sms.sendTextMessage(senderTel, null,
            "location:" + loc.getLatitude() + ":" +
            loc.getLongitude(), null, null);

        //---stop listening for location changes---
        lm.removeUpdates(locationListener);
    }
}
```

Note that once the location is found, you immediately remove the location listener.

Building the Location Tracker (Tablet)

Now that the location tracker for the smartphone is created, it is time to create the application that enables you to display a map showing the location of the person you are tracking. For this, you can use the original SMS application that you created earlier in this chapter. In the following exercise, you'll add the Google Maps `MapView` to the application so that it can be used to display the location of the person you are tracking.

TRY IT OUT   Adding Google Maps to the SMS Application

1.   Right-click on the sms project name in Eclipse and select Properties.

2.   Select Android on the left side of the Properties window and then select the Google APIs target on the right (see Figure 5-9). Click OK.

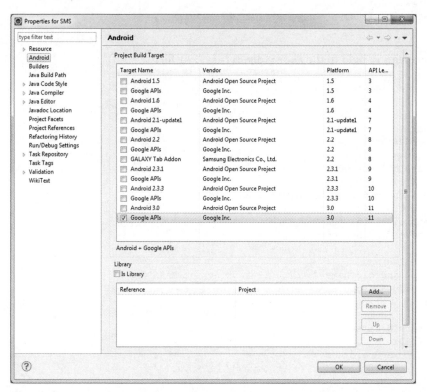

FIGURE 5-9

3.   Add the following statements in bold to the `main.xml` file (be sure to replace the Maps API key with your own):

```xml
<?xml version="1.0" encoding="utf-8"?>
<LinearLayout xmlns:android="http://schemas.android.com/apk/res/android"
    android:orientation="vertical"
    android:layout_width="fill_parent"
    android:layout_height="fill_parent" >
<Button
    android:id="@+id/btnSendSMS"
    android:layout_width="fill_parent"
    android:layout_height="wrap_content"
    android:text="Send SMS" />
<TextView
    android:id="@+id/textView1"
```

```
        android:layout_width="wrap_content"
        android:layout_height="wrap_content" />
<com.google.android.maps.MapView
    android:id="@+id/mapView"
    android:layout_width="fill_parent"
    android:layout_height="fill_parent"
    android:enabled="true"
    android:clickable="true"
    android:apiKey="0K2eMNyjc5HFPsiobLh6uLHb8F9ZFmh4uIm7VTA" />
</LinearLayout>
```

4. Add the following statements in bold to the `AndroidManifest.xml` file:

```
<?xml version="1.0" encoding="utf-8"?>
<manifest
    xmlns:android="http://schemas.android.com/apk/res/android"
    package="net.learn2develop.SMS"
    android:versionCode="1"
    android:versionName="1.0">
    <application android:icon="@drawable/icon"
    android:label="@string/app_name">
        <uses-library android:name="com.google.android.maps" />
        <activity android:name=".MainActivity"
                    android:label="@string/app_name"
                    android:launchMode="singleTask"
                    >
            <intent-filter>
                <action android:name="android.intent.action.MAIN" />
                <category android:name="android.intent.category.LAUNCHER" />
            </intent-filter>
        </activity>
        <receiver android:name=".SMSReceiver">
            <intent-filter>
                <action android:name=
                    "android.provider.Telephony.SMS_RECEIVED" />
            </intent-filter>
        </receiver>
    </application>
    <uses-sdk android:minSdkVersion="11" />
    <uses-permission android:name="android.permission.SEND_SMS">
    </uses-permission>
    <uses-permission android:name="android.permission.RECEIVE_SMS">
    </uses-permission>
    <uses-permission android:name="android.permission.INTERNET">
    </uses-permission>
</manifest>
```

5. Modify the `SMSReceiver.java` file as follows:

```java
package net.learn2develop.SMS;

import android.content.BroadcastReceiver;
import android.content.Context;
import android.content.Intent;
```

```java
import android.os.Bundle;
import android.telephony.SmsMessage;
import android.widget.Toast;

public class SMSReceiver extends BroadcastReceiver
{
    @Override
    public void onReceive(Context context, Intent intent)
    {
        //---get the SMS message passed in---
        Bundle bundle = intent.getExtras();
        SmsMessage[] msgs = null;
        String str = "";
        if (bundle != null)
        {
            //---retrieve the SMS message received---
            Object[] pdus = (Object[]) bundle.get("pdus");
            msgs = new SmsMessage[pdus.length];
            for (int i=0; i<msgs.length; i++){
                //---get the body of the message---
                msgs[i] = SmsMessage.createFromPdu((byte[])pdus[i]);
                str += msgs[i].getMessageBody().toString();
            }
            //---display the new SMS message---
            Toast.makeText(context, str, Toast.LENGTH_SHORT).show();

            //---launch the MainActivity---
            Intent mainActivityIntent = new
                Intent(context, MainActivity.class);
            mainActivityIntent.setFlags(Intent.FLAG_ACTIVITY_NEW_TASK);
            context.startActivity(mainActivityIntent);

            //---if the message body starts with location:---
            if (str.startsWith("location:")) {
                // e.g. location:1.23566:103.222344
                //---send a broadcast to update the SMS
                // received in the activity---
                Intent broadcastIntent = new Intent();
                broadcastIntent.setAction("SMS_RECEIVED_ACTION");
                broadcastIntent.putExtra("sms", str);
                context.sendBroadcast(broadcastIntent);
            }
        }
    }
}
```

6. Modify the `MainActivity.java` as follows:

```java
package net.learn2develop.SMS;

import android.content.BroadcastReceiver;
import android.content.Context;
import android.content.Intent;
```

```java
import android.content.IntentFilter;

import android.os.Bundle;
import android.view.View;
import android.widget.Button;
import android.widget.TextView;

import com.google.android.maps.GeoPoint;
import com.google.android.maps.MapActivity;
import com.google.android.maps.MapController;
import com.google.android.maps.MapView;

public class MainActivity extends MapActivity {
    Button btnSendSMS;
    IntentFilter intentFilter;

    private MapView mapView;
    private MapController mc;

    private BroadcastReceiver intentReceiver = new BroadcastReceiver() {
        @Override
        public void onReceive(Context context, Intent intent) {
            //---display the SMS received in the TextView---
            TextView SMSes = (TextView) findViewById(R.id.textView1);
            SMSes.setText(intent.getExtras().getString("sms"));

            //---Make the map display the location information received---
            // e.g. location:1.23566:103.222344
            String[] coordinates =
                intent.getExtras().getString("sms").split(":");
            double lat = Double.parseDouble(coordinates[1]);
            double lng = Double.parseDouble(coordinates[2]);
            GeoPoint p = new GeoPoint((int) (lat * 1E6),
                             (int) (lng * 1E6));
            mc.animateTo(p);
            mc.setZoom(16);
        }
    };

    /** Called when the activity is first created. */
    @Override
    public void onCreate(Bundle savedInstanceState) {
        super.onCreate(savedInstanceState);
        setContentView(R.layout.main);

        mapView = (MapView) findViewById(R.id.mapView);
        mc = mapView.getController();

        //---intent to filter for SMS messages received---
        intentFilter = new IntentFilter();
        intentFilter.addAction("SMS_RECEIVED_ACTION");

        //---register the receiver---
```

```
        registerReceiver(intentReceiver, intentFilter);

        btnSendSMS = (Button) findViewById(R.id.btnSendSMS);
        btnSendSMS.setOnClickListener(new View.OnClickListener()
        {
            public void onClick(View v)
            {
                //sendSMS("5554", "Hello my friends!");
                Intent i = new
                    Intent(android.content.Intent.ACTION_VIEW);
                i.putExtra("address", "5556");
                i.putExtra("sms_body", "Where are you?");
                i.setType("vnd.android-dir/mms-sms");
                startActivity(i);
            }
        });
    }

    @Override
    protected void onResume() {
        //---register the receiver---
        //registerReceiver(intentReceiver, intentFilter);
        super.onResume();
    }

    @Override
    protected void onPause() {
        //---unregister the receiver---
        //unregisterReceiver(intentReceiver);
        super.onPause();
    }

    @Override
    protected void onDestroy() {
        //---unregister the receiver---
        unregisterReceiver(intentReceiver);
        super.onPause();
    }

    @Override
    protected boolean isRouteDisplayed() {
        // TODO Auto-generated method stub
        return false;
    }
}
```

Alright! After the How It Works explanation, you learn how to test the two applications.

How It Works

For this application, you added the Google Maps MapView to the UI of the application. For ease of testing the application on the Android emulator, you hard-coded the Send SMS button to send an SMS message to 5556, with the text "Where are you?":

```
btnSendSMS = (Button) findViewById(R.id.btnSendSMS);
btnSendSMS.setOnClickListener(new View.OnClickListener()
{
    public void onClick(View v)
    {
        //sendSMS("5554", "Hello my friends!");
        Intent i = new
            Intent(android.content.Intent.ACTION_VIEW);
        i.putExtra("address", "5556");
        i.putExtra("sms_body", "Where are you?");
        i.setType("vnd.android-dir/mms-sms");
        startActivity(i);
    }
});
```

When the location tracker application returns the SMS message containing the location information (starting with the word "location:"), you broadcast an intent containing the content of the SMS message:

```
//---retrieve the SMS message received---
Object[] pdus = (Object[]) bundle.get("pdus");
msgs = new SmsMessage[pdus.length];
for (int i=0; i<msgs.length; i++){
    //---get the body of the message---
    msgs[i] = SmsMessage.createFromPdu((byte[])pdus[i]);
    str += msgs[i].getMessageBody().toString();
}
//---display the new SMS message---
Toast.makeText(context, str, Toast.LENGTH_SHORT).show();

//---launch the MainActivity---
Intent mainActivityIntent = new
    Intent(context, MainActivity.class);
mainActivityIntent.setFlags(Intent.FLAG_ACTIVITY_NEW_TASK);
context.startActivity(mainActivityIntent);

if (str.startsWith("location:")) {
    // e.g. location:1.23566:103.222344
    //---send a broadcast to update the SMS
    // received in the activity---
    Intent broadcastIntent = new Intent();
    broadcastIntent.setAction("SMS_RECEIVED_ACTION");
    broadcastIntent.putExtra("sms", str);
    context.sendBroadcast(broadcastIntent);
}
```

In `MainActivity`, you extract the latitude and longitude from the SMS message and then navigate the `MapView` to display the location of the user:

```
//---Make the map display the location information received---
// e.g. location:1.23566:103.222344
String[] coordinates =
    intent.getExtras().getString("sms").split(":");
double lat = Double.parseDouble(coordinates[1]);
```

```
double lng = Double.parseDouble(coordinates[2]);
GeoPoint p = new GeoPoint((int) (lat * 1E6),
                (int) (lng * 1E6));
mc.animateTo(p);
mc.setZoom(16);
```

Testing the Applications

To test the two applications, now launch two Android emulators. For consistency, you will first launch an Android 3.0 tablet emulator (using the Google APIs) with the port number 5554 (henceforth referred to as the Android 5554 emulator). Then, launch another Android 2.2 emulator with port number 5556 (henceforth referred to as the Android 5556 emulator).

In Eclipse, deploy the SMS tablet application onto the Android 5554 emulator, and the LocationTracker application onto the Android 5556 emulator.

On the Android tablet application, click the Send SMS button (see Figure 5-10).

FIGURE 5-10

The Messaging application launches. Because you have hard-coded the recipient and content of the message to send, simply click the Send button to send the message (see Figure 5-11). Note that on the Android 3.0 emulator, the content of the message is not shown.

FIGURE 5-11

Once the SMS is sent, it should be received by the Android 5556 emulator (see Figure 5-12).

FIGURE 5-12

To simulate the Android 5556 emulator having a location fix (this is the term that refers to the process of a device obtaining a location determination from GPS) returned by its GPS receiver, you will use the DDMS perspective in Eclipse to send a pair of coordinates to it. Figure 5-13 shows that you first have to select the Android 5556 emulator, and then click the Send button to send the latitude and longitude to the selected emulator.

Once the Android 5556 emulator receives the location coordinates, it will send an SMS message to the Android 5554 emulator. Figure 5-14 shows the Android 5554 emulator receiving the SMS message containing the location data, and Google Maps navigating to display the location.

Have fun!

FIGURE 5-13

FIGURE 5-14

SENDING E-MAIL

Like SMS messaging, Android also supports e-mail. The Gmail/Email application on Android enables you to configure an e-mail account using POP3 or IMAP. Besides sending and receiving e-mails using the Gmail/Email application, you can also send e-mail messages programmatically from within your Android application. The following Try It Out shows you how.

Sending E-Mail Programmatically

codefile Emails.zip available for download at Wrox.com

1. Using Eclipse, create a new Android project and name it `Emails`.

2. Add the following statements in bold to the `main.xml` file:

```xml
<?xml version="1.0" encoding="utf-8"?>
<LinearLayout xmlns:android="http://schemas.android.com/apk/res/android"
    android:orientation="vertical"
    android:layout_width="fill_parent"
    android:layout_height="fill_parent" >
<Button
    android:id="@+id/btnSendEmail"
    android:layout_width="fill_parent"
    android:layout_height="wrap_content"
    android:text="Send Email" />
</LinearLayout>
```

3. Add the following statements in bold to the `MainActivity.java` file:

```java
package net.learn2develop.Email;

import android.app.Activity;
import android.os.Bundle;

import android.content.Intent;
import android.net.Uri;
import android.view.View;
import android.widget.Button;

public class MainActivity extends Activity {
    Button btnSendEmail;

    /** Called when the activity is first created. */
    @Override
    public void onCreate(Bundle savedInstanceState) {
        super.onCreate(savedInstanceState);
        setContentView(R.layout.main);

        btnSendEmail = (Button) findViewById(R.id.btnSendEmail);
        btnSendEmail.setOnClickListener(new View.OnClickListener()
        {
            public void onClick(View v)
            {
                String[] to = {"weimenglee@learn2develop.net",
                               "weimenglee@gmail.com"};
                String[] cc = {"course@learn2develop.net"};
                sendEmail(to, cc, "Hello", "Hello my friends!");
            }
        });
```

```
    }

    //---sends an SMS message to another device---
    private void sendEmail(String[] emailAddresses, String[] carbonCopies,
    String subject, String message)
    {
        Intent emailIntent = new Intent(Intent.ACTION_SEND);
        emailIntent.setData(Uri.parse("mailto:"));
        String[] to = emailAddresses;
        String[] cc = carbonCopies;
        emailIntent.putExtra(Intent.EXTRA_EMAIL, to);
        emailIntent.putExtra(Intent.EXTRA_CC, cc);
        emailIntent.putExtra(Intent.EXTRA_SUBJECT, subject);
        emailIntent.putExtra(Intent.EXTRA_TEXT, message);
        emailIntent.setType("message/rfc822");
        startActivity(Intent.createChooser(emailIntent, "Email"));
    }
}
```

4. Press F11 to test the application on a real Android device. Click the Send Email button and you should see the Email application launched in your device. Note that if you test this application on the Android emulator (both 2.2 and 3.0), it will display a "No applications can perform this action" message.

How It Works

In this example, you launched the built-in Email application to send an e-mail message. To do so, you used an `Intent` object, setting the various parameters using the `setData()`, `putExtra()`, and `setType()` methods:

```
Intent emailIntent = new Intent(Intent.ACTION_SEND);
emailIntent.setData(Uri.parse("mailto:"));
String[] to = emailAddresses;
String[] cc = carbonCopies;
emailIntent.putExtra(Intent.EXTRA_EMAIL, to);
emailIntent.putExtra(Intent.EXTRA_CC, cc);
emailIntent.putExtra(Intent.EXTRA_SUBJECT, subject);
emailIntent.putExtra(Intent.EXTRA_TEXT, message);
emailIntent.setType("message/rfc822");
startActivity(Intent.createChooser(emailIntent, "Email"));
```

NETWORKING

The previous sections covered how to connect to the outside world using SMS and e-mail. Another way to achieve that is to use the HTTP protocol. Using the HTTP protocol, you can perform a wide variety of tasks, such as downloading web pages from a web server, downloading binary data, and so on.

The following Try It Out creates an Android project so you can use the HTTP protocol to connect to the Web to download all sorts of data.

TRY IT OUT Creating the Base Project for HTTP Connection

codefile Networking.zip available for download at Wrox.com

1. Using Eclipse, create a new Android 3.0 project and name it **Networking**.

2. Add the following statement in bold to the `AndroidManifest.xml` file:

```xml
<?xml version="1.0" encoding="utf-8"?>
<manifest xmlns:android="http://schemas.android.com/apk/res/android"
      package="net.learn2develop.Networking"
      android:versionCode="1"
      android:versionName="1.0">
    <application android:icon="@drawable/icon" android:label="@string/app_name">
        <activity android:name=".MainActivity"
                  android:label="@string/app_name">
            <intent-filter>
                <action android:name="android.intent.action.MAIN" />
                <category android:name="android.intent.category.LAUNCHER" />
            </intent-filter>
        </activity>
    </application>
    <uses-sdk android:minSdkVersion="8" />
    <uses-permission android:name="android.permission.INTERNET"></uses-permission>
</manifest>
```

3. Import the following namespaces in the `MainActivity.java` file:

```java
package net.learn2develop.Networking;

import android.app.Activity;
import android.os.Bundle;

import java.io.IOException;
import java.io.InputStream;
import java.io.InputStreamReader;
import java.net.HttpURLConnection;
import java.net.URL;
import java.net.URLConnection;
import android.graphics.Bitmap;
import android.graphics.BitmapFactory;
import android.widget.ImageView;
import android.widget.Toast;

import javax.xml.parsers.DocumentBuilder;
import javax.xml.parsers.DocumentBuilderFactory;
import javax.xml.parsers.ParserConfigurationException;

import org.w3c.dom.Document;
import org.w3c.dom.Element;
import org.w3c.dom.Node;
```

```
import org.w3c.dom.NodeList;

public class MainActivity extends Activity {
    /** Called when the activity is first created. */
    @Override
    public void onCreate(Bundle savedInstanceState) {
        super.onCreate(savedInstanceState);
        setContentView(R.layout.main);
    }
}
```

4. Define the OpenHttpConnection() method in the MainActivity.java file:

```
public class MainActivity extends Activity {

    private InputStream OpenHttpConnection(String urlString)
    throws IOException
    {
        InputStream in = null;
        int response = -1;

        URL url = new URL(urlString);
        URLConnection conn = url.openConnection();

        if (!(conn instanceof HttpURLConnection))
            throw new IOException("Not an HTTP connection");
        try{
            HttpURLConnection httpConn = (HttpURLConnection) conn;
            httpConn.setAllowUserInteraction(false);
            httpConn.setInstanceFollowRedirects(true);
            httpConn.setRequestMethod("GET");
            httpConn.connect();
            response = httpConn.getResponseCode();
            if (response == HttpURLConnection.HTTP_OK) {
                in = httpConn.getInputStream();
            }
        }
        catch (Exception ex)
        {
            throw new IOException("Error connecting");
        }
        return in;
    }

    /** Called when the activity is first created. */
    @Override
    public void onCreate(Bundle savedInstanceState) {
        super.onCreate(savedInstanceState);
        setContentView(R.layout.main);
    }
}
```

How It Works

Because you are using the HTTP protocol to connect to the Web, your application needs the INTERNET permission; hence, the first thing you did is add the permission in the AndroidManifest.xml file.

You then defined the OpenHttpConnection() method, which takes a URL string and returns an InputStream object. Using an InputStream object, you can download the data by reading bytes from the stream object. In this method, you made use of the HttpURLConnection object to open an HTTP connection with a remote URL. You set all the various properties of the connection, such as the request method, and so on:

```
HttpURLConnection httpConn = (HttpURLConnection) conn;
httpConn.setAllowUserInteraction(false);
httpConn.setInstanceFollowRedirects(true);
httpConn.setRequestMethod("GET");
```

After you try to establish a connection with the server, you get the HTTP response code from it. If the connection is established (via the response code HTTP_OK), then you proceed to get an InputStream object from the connection:

```
httpConn.connect();
response = httpConn.getResponseCode();
if (response == HttpURLConnection.HTTP_OK) {
    in = httpConn.getInputStream();
}
```

Using the InputStream object, you can then start to download the data from the server.

Downloading Binary Data

One of the common tasks you need to perform is downloading binary data from the Web. For example, you may want to download an image from a server so that you can display it in your application. The following Try It Out shows how this is done.

TRY IT OUT Downloading Binary Data

1. Using the same project created earlier, add the following statements in bold to the main .xml file:

```
<?xml version="1.0" encoding="utf-8"?>
<LinearLayout xmlns:android="http://schemas.android.com/apk/res/android"
    android:orientation="vertical"
    android:layout_width="fill_parent"
    android:layout_height="fill_parent" >
<ImageView
    android:id="@+id/img"
```

```
      android:layout_width="wrap_content"
      android:layout_height="wrap_content"
      android:layout_gravity="center" />
</LinearLayout>
```

2. Add the following statements in bold to the MainActivity.java file:

```
public class MainActivity extends Activity {
    ImageView img;

    private InputStream OpenHttpConnection(String urlString)
    throws IOException
    {
        //...
    }

    private Bitmap DownloadImage(String URL)
    {
        Bitmap bitmap = null;
        InputStream in = null;
        try {
            in = OpenHttpConnection(URL);
            bitmap = BitmapFactory.decodeStream(in);
            in.close();
        } catch (IOException e1) {
            Toast.makeText(this, e1.getLocalizedMessage(),
                Toast.LENGTH_LONG).show();

            e1.printStackTrace();
        }
        return bitmap;
    }

    /** Called when the activity is first created. */
    @Override
    public void onCreate(Bundle savedInstanceState) {
        super.onCreate(savedInstanceState);
        setContentView(R.layout.main);

        //---download an image---
        Bitmap bitmap =
            DownloadImage(
            "http://www.mayoff.com/5-01cablecarDCP01934.jpg");
        img = (ImageView) findViewById(R.id.img);
        img.setImageBitmap(bitmap);
    }
}
```

3. Press F11 to debug the application on the Android emulator. Figure 5-15 shows the image downloaded from the Web and then displayed in the ImageView.

FIGURE 5-15

How It Works

The `DownloadImage()` method takes the URL of the image to download and then opens the connection to the server using the `OpenHttpConnection()` method that you have defined earlier. Using the `InputStream` object returned by the connection, the `decodeStream()` method from the `BitmapFactory` class is used to download and decode the data into a `Bitmap` object. The `DownloadImage()` method returns a `Bitmap` object.

The image is then displayed using an ImageView view.

REFERRING TO LOCALHOST FROM YOUR EMULATOR

When working with the Android emulator, you may frequently need to access data hosted on the local web server using `localhost`. For example, your own Web services are likely to be hosted on your local computer during development, and you'll want to test them on the same development machine you use to write your Android applications. In such cases, you should use the special IP address of 10.0.2.2 (not 127.0.0.1) to refer to the host computer's loopback interface. From the Android emulator's perspective, `localhost` (127.0.0.1) refers to its own loopback interface.

Downloading Text Files

Besides downloading binary data, you can also download plain-text files. For example, you might be writing an RSS Reader application and therefore need to download RSS XML feeds for processing. The following Try It Out shows how you can download a plain-text file in your application.

TRY IT OUT Downloading Plain-Text Files

1. Using the same project created earlier, add the following statements in bold to the MainActivity .java file:

```java
public class MainActivity extends Activity {
    ImageView img;

    private InputStream OpenHttpConnection(String urlString)
    throws IOException
    {
        //...
    }

    private Bitmap DownloadImage(String URL)
    {
        //...
    }

    private String DownloadText(String URL)
    {
        int BUFFER_SIZE = 2000;
        InputStream in = null;
        try {
            in = OpenHttpConnection(URL);
        } catch (IOException e1) {
            Toast.makeText(this, e1.getLocalizedMessage(),
                Toast.LENGTH_LONG).show();

            e1.printStackTrace();
            return "";
        }

        InputStreamReader isr = new InputStreamReader(in);
        int charRead;
        String str = "";
        char[] inputBuffer = new char[BUFFER_SIZE];
        try {
            while ((charRead = isr.read(inputBuffer))>0)
            {
                //---convert the chars to a String---
                String readString =
                    String.copyValueOf(inputBuffer, 0, charRead);
                str += readString;
                inputBuffer = new char[BUFFER_SIZE];
            }
            in.close();
        } catch (IOException e) {
            Toast.makeText(this, e.getLocalizedMessage(),
```

```
                        Toast.LENGTH_LONG).show();

            e.printStackTrace();
            return "";
        }
        return str;
    }


    /** Called when the activity is first created. */
    @Override
    public void onCreate(Bundle savedInstanceState) {
        super.onCreate(savedInstanceState);
        setContentView(R.layout.main);

        //---download an image---
        Bitmap bitmap =
            DownloadImage(
            "http://www.streetcar.org/mim/cable/images/cable-01.jpg");
        img = (ImageView) findViewById(R.id.img);
        img.setImageBitmap(bitmap);

        //---download an RSS feed---
        String str = DownloadText(
            "http://www.appleinsider.com/appleinsider.rss");
        Toast.makeText(getBaseContext(), str,
                        Toast.LENGTH_SHORT).show();
    }
}
```

2. Press F11 to debug the application on the Android emulator. Figure 5-16 shows the RSS feed
    downloaded and displayed using the Toast class.

FIGURE 5-16

How It Works

The `DownloadText()` method takes an URL of the text file to download and then returns the string of the text file downloaded. It basically opens an HTTP connection to the server and then uses an `InputStreamReader` object to read each character from the stream and save it in a `String` object.

Accessing Web Services Using the GET Method

So far, you have learned how to download images and text from the Web. The previous section showed how to download an RSS feed from a server. Very often, you need to download XML files and parse the contents (a good example of this is consuming Web services). Therefore, in this section you learn how to connect to a Web service using the HTTP GET method. Once the Web service returns a result in XML, you will extract the relevant parts and display its content using the `Toast` class.

For this example, the web method you will be using is from `http://services.aonaware.com/DictService/DictService.asmx?op=Define`. This web method is from a Dictionary Web service that returns the definition of a given word.

The web method takes a request in the following format:

```
GET /DictService/DictService.asmx/Define?word=string HTTP/1.1
Host: services.aonaware.com
HTTP/1.1 200 OK
Content-Type: text/xml; charset=utf-8
Content-Length: length
```

It returns a response in the following format:

```
<?xml version="1.0" encoding="utf-8"?>
<WordDefinition xmlns="http://services.aonaware.com/webservices/">
  <Word>string</Word>
  <Definitions>
    <Definition>
      <Word>string</Word>
      <Dictionary>
        <Id>string</Id>
        <Name>string</Name>
      </Dictionary>
      <WordDefinition>string</WordDefinition>
    </Definition>
    <Definition>
      <Word>string</Word>
      <Dictionary>
        <Id>string</Id>
        <Name>string</Name>
      </Dictionary>
      <WordDefinition>string</WordDefinition>
    </Definition>
  </Definitions>
</WordDefinition>
```

Hence, to obtain the definition of a word, you need to establish an HTTP connection to the web method and then parse the XML result that is returned. The following Try It Out shows you how.

TRY IT OUT Consuming Web Services

1. Using the same project created earlier, add the following statements in bold to the `MainActivity` `.java` file:

```java
public class MainActivity extends Activity {
    ImageView img;

    private InputStream OpenHttpConnection(String urlString)
    throws IOException
    {
        //...
    }

    private Bitmap DownloadImage(String URL)
    {
        //...
    }

    private String DownloadText(String URL)
    {
        //...
    }

    private void WordDefinition(String word) {
        InputStream in = null;
        try {
            in = OpenHttpConnection(
    "http://services.aonaware.com/DictService/DictService.asmx/Define?word=" + word);
            Document doc = null;
            DocumentBuilderFactory dbf =
                DocumentBuilderFactory.newInstance();
            DocumentBuilder db;
            try {
                db = dbf.newDocumentBuilder();
                doc = db.parse(in);
            } catch (ParserConfigurationException e) {
                // TODO Auto-generated catch block
                e.printStackTrace();
            } catch (Exception e) {
                // TODO Auto-generated catch block
                e.printStackTrace();
            }
            doc.getDocumentElement().normalize();

            //---retrieve all the <Definition> nodes---
            NodeList itemNodes =
                doc.getElementsByTagName("Definition");

            String strDefinition = "";
```

```java
            for (int i = 0; i < definitionElements.getLength(); i++) {
                Node itemNode = definitionElements.item(i);
                if (itemNode.getNodeType() == Node.ELEMENT_NODE)
                {
                    //---convert the Node into an Element---
                    Element definitionElement = (Element) itemNode;

                    //---get all the <WordDefinition> elements under
                    // the <Definition> element---
                    NodeList wordDefinitionElements =
                        (definitionElement).getElementsByTagName(
                        "WordDefinition");

                    strDefinition = "";
                    for (int j = 0; j < wordDefinitionElements.getLength(); j++) {
                        //---convert a <WordDefinition> Node into an Element---
                        Element wordDefinitionElement =
                            (Element) wordDefinitionElements.item(j);

                        //---get all the child nodes under the
                        // <WordDefinition> element---
                        NodeList textNodes =
                            ((Node) wordDefinitionElement).getChildNodes();

                        strDefinition +=
                            ((Node) textNodes.item(0)).getNodeValue() + ". ";
                    }

                    //---display the title---
                    Toast.makeText(getBaseContext(),strDefinition,
                        Toast.LENGTH_SHORT).show();
                }
            }
        } catch (IOException e1) {
            Toast.makeText(this, e1.getLocalizedMessage(),
                Toast.LENGTH_LONG).show();
            e1.printStackTrace();
        }
    }

    /** Called when the activity is first created. */
    @Override
    public void onCreate(Bundle savedInstanceState) {
        super.onCreate(savedInstanceState);
        setContentView(R.layout.main);

        //---download an image---
        Bitmap bitmap =
            DownloadImage(
            "http://www.mayoff.com/5-01cablecarDCP01934.jpg");
        img = (ImageView) findViewById(R.id.img);
        img.setImageBitmap(bitmap);

        //---download an RSS feed---
```

```
        String str = DownloadText(
            "http://www.appleinsider.com/appleinsider.rss");
        Toast.makeText(getBaseContext(), str,
                        Toast.LENGTH_SHORT).show();

        //---access a Web service using GET---
        WordDefinition("Apple");
    }
}
```

2. Press F11 to debug the application on the Android emulator. Figure 5-17 shows the result of the Web service call being parsed and then displayed using the Toast class.

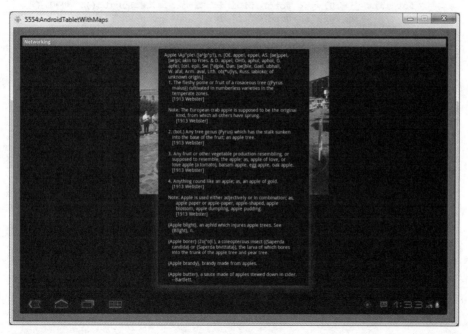

FIGURE 5-17

How It Works

The WordDefinition() method first opens an HTTP connection to the Web service, passing in the word that you are interested in:

```
            in = OpenHttpConnection(
        "http://services.aonaware.com/DictService/DictService.asmx/Define?word=" + word);
```

It then uses the DocumentBuilderFactory and DocumentBuilder objects to obtain a Document (DOM) object from an XML file (which is the XML result returned by the Web service):

```
Document doc = null;
DocumentBuilderFactory dbf =
    DocumentBuilderFactory.newInstance();
DocumentBuilder db;
try {
    db = dbf.newDocumentBuilder();
    doc = db.parse(in);
} catch (ParserConfigurationException e) {
    // TODO Auto-generated catch block
    e.printStackTrace();
} catch (Exception e) {
    // TODO Auto-generated catch block
    e.printStackTrace();
}
doc.getDocumentElement().normalize();
```

Once the Document object is obtained, you find all the elements with the <Definition> tag:

```
//---retrieve all the <Definition> nodes---
NodeList itemNodes =
    doc.getElementsByTagName("Definition");
```

Figure 5-18 shows the structure of the XML document returned by the Web service.

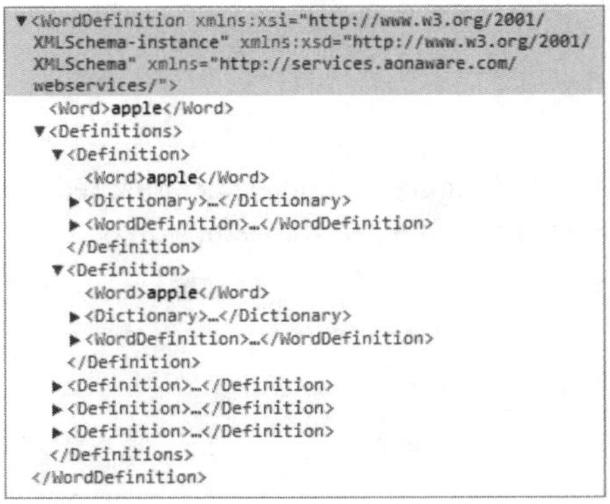

```
▼<WordDefinition xmlns:xsi="http://www.w3.org/2001/
  XMLSchema-instance" xmlns:xsd="http://www.w3.org/2001/
  XMLSchema" xmlns="http://services.aonaware.com/
  webservices/">
  <Word>apple</Word>
  ▼<Definitions>
    ▼<Definition>
        <Word>apple</Word>
      ▶<Dictionary>…</Dictionary>
      ▶<WordDefinition>…</WordDefinition>
      </Definition>
    ▼<Definition>
        <Word>apple</Word>
      ▶<Dictionary>…</Dictionary>
      ▶<WordDefinition>…</WordDefinition>
      </Definition>
    ▶<Definition>…</Definition>
    ▶<Definition>…</Definition>
    ▶<Definition>…</Definition>
    </Definitions>
  </WordDefinition>
```

FIGURE 5-18

Because the definition of a word is contained within the <WordDefinition> element, you then proceed to extract all the definitions:

```
String strDefinition = "";
for (int i = 0; i < definitionElements.getLength(); i++) {
    Node itemNode = definitionElements.item(i);
    if (itemNode.getNodeType() == Node.ELEMENT_NODE)
    {
        //---convert the Node into an Element---
        Element definitionElement = (Element) itemNode;

        //---get all the <WordDefinition> elements under
```

```
                    // the <Definition> element---
                    NodeList wordDefinitionElements =
                        (definitionElement).getElementsByTagName(
                        "WordDefinition");

                    strDefinition = "";
                    for (int j = 0; j < wordDefinitionElements.getLength(); j++) {
                        //---convert a <WordDefinition> Node into an Element---
                        Element wordDefinitionElement =
                            (Element) wordDefinitionElements.item(j);

                        //---get all the child nodes under the
                        // <WordDefinition> element---
                        NodeList textNodes =
                            ((Node) wordDefinitionElement).getChildNodes();
                        //---get the first node, which contains the text---
                        strDefinition +=
                            ((Node) textNodes.item(0)).getNodeValue() + ". ";
                    }
                    //---display the title---
                    Toast.makeText(getBaseContext(),strDefinition,
                        Toast.LENGTH_SHORT).show();
                }
            }
        } catch (IOException e1) {
            Toast.makeText(this, e1.getLocalizedMessage(),
                Toast.LENGTH_LONG).show();
            e1.printStackTrace();
        }
```

The preceding code loops through all the <Definition> elements looking for a child element named <WordDefinition>. The text content of the <WordDefinition> element contains the definition of a word. The Toast class displays each word definition that is retrieved.

Performing Asynchronous Calls

All the connections made in the previous few sections have been *synchronous* — that is, the connection to a server will not return until the data is received. In real life, this presents some problems due to network connections being inherently slow. When you connect to a server to download some data, the user interface of your application remains frozen until a response is obtained. In most cases, this is not acceptable. Hence, you need to ensure that the connection to the server is made in an asynchronous fashion.

The easiest way to connect to the server asynchronously is to use the AsyncTask class available in the Android SDK. Using AsyncTask enables you to perform background tasks in a separate thread and then return the result in a UI thread. That way, you can perform background operations without needing to handle complex threading issues.

Using the previous example of downloading an image from the server and then displaying the image in an `ImageView`, you could wrap the code in an instance of the `AsyncTask` class, as shown here:

```
public class MainActivity extends Activity {
    ImageView img;

    private class BackgroundTask extends AsyncTask
    <String, Void, Bitmap> {
        protected Bitmap doInBackground(String... url) {
            //---download an image---
            Bitmap bitmap = DownloadImage(url[0]);
            return bitmap;
        }

        protected void onPostExecute(Bitmap bitmap) {
            ImageView img = (ImageView) findViewById(R.id.img);
            img.setImageBitmap(bitmap);
        }
    }

    private InputStream OpenHttpConnection(String urlString)
    throws IOException
    {
        ...
    }
}
```

Basically, you defined a class that extends the `AsyncTask` class. In this case, there are two methods within the `BackgroundTask` class: `doInBackground()` and `onPostExecute()`. You put all the code that needs to be run asynchronously in the `doInBackground()` method. When the task is completed, the result is passed back via the `onPostExecute()` method. The `onPostExecute()` method is executed on the UI thread, hence it is thread safe to update the `ImageView` with the bitmap downloaded from the server.

To perform the asynchronous tasks, simply create an instance of the `BackgroundTask` class and call its `execute()` method:

```
@Override
public void onCreate(Bundle savedInstanceState) {
    super.onCreate(savedInstanceState);
    setContentView(R.layout.main);
        new BackgroundTask().execute(
            "http://www.mayoff.com/5-01cablecarDCP01934.jpg");
}
```

SUMMARY

This chapter described the various ways to communicate with the outside world. You first learned how to send and receive SMS messages. You then learned how to send e-mail messages from within your Android application. The chapter ended with lessons on using the HTTP protocol to download data from a web server.

EXERCISES

1. Name the two ways in which you can send SMS messages in your Android application.

2. Name the permissions you need to declare in your `AndroidManifest.xml` file for sending and receiving SMS messages.

3. How do you notify an activity from a `BroadcastReceiver`?

4. Name the permissions you need to declare in your `AndroidManifest.xml` file for an HTTP connection.

Answers to the Exercises can be found in Appendix C.

▶ WHAT YOU LEARNED IN THIS CHAPTER

TOPIC	KEY CONCEPTS
Programmatically sending SMS messages	Use the `SmsManager` class.
Getting feedback on messages sent	Use two `PendingIntent` objects in the `sendTextMessage()` method.
Sending SMS messages using Intent	Set the intent type to "`vnd.android-dir/mms-sms`."
Receiving SMS messages	Implement a `BroadcastReceiver` and set it in the `AndroidManifest.xml` file.
Sending e-mail using Intent	Set the intent type to "`message/rfc822`."
Establishing an HTTP connection	Use the `HttpURLConnection` class.
Accessing Web services	Use the `Document`, `DocumentBuilderFactory`, and `DocumentBuilder` classes to parse the XML result returned by the Web service.

Publishing Android Applications

WHAT YOU WILL LEARN IN THIS CHAPTER

➤ How to prepare your application for deployment

➤ How to export your application as an APK file and sign it with a new certificate

➤ How to distribute your Android application

➤ How to publish your application on the Android Market

So far you have seen quite a lot of interesting things you can do with your Android tablet. However, in order to get your application running on users' devices, you need a way to deploy it and distribute it. In this chapter, you will learn how to prepare your Android applications for deployment and get them onto your customer's devices. In addition, you will learn how to publish your applications on the Android Market, where you can sell them and make some money!

PREPARING FOR PUBLISHING

Google has made it relatively easy to publish your Android application so that it can be quickly distributed to end users. The steps to publishing your Android application generally involve the following:

1. Export your application as an APK (Android Package) file.

2. Generate your own self-signed certificate and digitally sign your application with it.

3. Deploy the signed application.

4. Use the Android Market for hosting and selling your application.

In the following sections, you will learn how to prepare your application for signing, and then learn about the various ways to deploy your applications.

This chapter uses the LBS project created in Chapter 4 to demonstrate how to deploy an Android application.

Versioning

Beginning with version 1.0 of the Android SDK, the AndroidManifest.xml file of every Android application includes the android:versionCode and android:versionName attributes:

```xml
<?xml version="1.0" encoding="utf-8"?>
<manifest xmlns:android="http://schemas.android.com/apk/res/android"
    package="net.learn2develop.LBS"
    android:versionCode="1"
    android:versionName="1.0">
    <application android:icon="@drawable/icon" android:label="@string/app_name">
    <uses-library android:name="com.google.android.maps" />
        <activity android:name=".MainActivity"
                android:label="@string/app_name">
            <intent-filter>
                <action android:name="android.intent.action.MAIN" />
                <category android:name="android.intent.category.LAUNCHER" />
            </intent-filter>
        </activity>
    </application>
    <uses-sdk android:minSdkVersion="11" />
        <uses-permission android:name="android.permission.INTERNET" />
<uses-permission android:name="android.permission.ACCESS_FINE_LOCATION" />
<uses-permission android:name="android.permission.ACCESS_COARSE_LOCATION" />
</manifest>
```

The android:versionCode attribute represents the version number of your application. For every revision you make to the application, you should increment this value by 1 so that you can programmatically differentiate the newest version from the previous one. This value is never used by the Android system, but it is useful for developers as a means to obtain the version number of an application. However, the android:versionCode attribute is used by Android Market to determine if a newer version of your application is available.

You can programmatically retrieve the value of the android:versionCode attribute by using the getPackageInfo() method from the PackageManager class, like this:

```java
PackageManager pm = getPackageManager();
try {
    //---get the package info---
    PackageInfo pi =
        pm.getPackageInfo("net.learn2develop.LBS", 0);
    //---display the versioncode---
    Toast.makeText(getBaseContext(),
        "VersionCode: " +Integer.toString(pi.versionCode),
        Toast.LENGTH_SHORT).show();
} catch (NameNotFoundException e) {
    // TODO Auto-generated catch block
    e.printStackTrace();
}
```

The `android:versionName` attribute contains versioning information that is visible to the users. It should contain values in the format: *<major>.<minor>.<point>*. If your application undergoes a major upgrade, you should increase the *<major>* by 1. For small incremental updates, you can increase either the *<minor>* or *<point>* by 1. For example, a new application may have a version name of "1.0.0." For a small incremental update, you might change it to "1.1.0" or "1.0.1." For the next major update, you might change it to "2.0.0."

If you are planning to publish your application on the Android Market (`www.android.com/market/`), the `AndroidManifest.xml` file must have the following attributes:

➤ `android:versionCode` (within the `<manifest>` element)

➤ `android:versionName` (within the `<manifest>` element)

➤ `android:icon` (within the `<application>` element)

➤ `android:label` (within the `<application>` element)

The `android:label` attribute specifies the name of your application. This name is displayed in the Settings ➪ Applications ➪ Manage Applications section of your Android device. For the LBS project, give the application the name "Where Am I":

```
<?xml version="1.0" encoding="utf-8"?>
<manifest xmlns:android="http://schemas.android.com/apk/res/android"
    package="net.learn2develop.LBS"
    android:versionCode="1"
    android:versionName="1.0">
  <application android:icon="@drawable/icon" android:label="Where Am I">
  <uses-library android:name="com.google.android.maps" />
    <activity android:name=".MainActivity"
            android:label="@string/app_name">
      <intent-filter>
        <action android:name="android.intent.action.MAIN" />
        <category android:name="android.intent.category.LAUNCHER" />
      </intent-filter>
    </activity>
  </application>
  <uses-sdk android:minSdkVersion="11" />
  <uses-permission android:name="android.permission.INTERNET" />
<uses-permission android:name="android.permission.ACCESS_FINE_LOCATION" />
<uses-permission android:name="android.permission.ACCESS_COARSE_LOCATION" />
</manifest>
```

In addition, if your application needs a minimum version of the SDK, you can specify it in the `AndroidManifest.xml` file using the `<uses-sdk>` element:

```
<?xml version="1.0" encoding="utf-8"?>
<manifest xmlns:android="http://schemas.android.com/apk/res/android"
    package="net.learn2develop.LBS"
    android:versionCode="1"
    android:versionName="1.0">
  <application android:icon="@drawable/icon" android:label="Where Am I">
```

```
        <uses-library android:name="com.google.android.maps" />

            <activity android:name=".MainActivity"
                    android:label="@string/app_name">
                <intent-filter>
                    <action android:name="android.intent.action.MAIN" />
                    <category android:name="android.intent.category.LAUNCHER" />
                </intent-filter>
            </activity>
        </application>
        <uses-sdk android:minSdkVersion="11" />
        <uses-permission android:name="android.permission.INTERNET" />
        <uses-permission android:name="android.permission.ACCESS_FINE_LOCATION" />
        <uses-permission android:name="android.permission.ACCESS_COARSE_LOCATION" />
    </manifest>
```

In the preceding example, the application requires a minimum of SDK version 11, which is Android 3.0. In general, it is always good to set this version number to the lowest one that your application can support. This ensures that a wider range of users will be able to run your application.

Digitally Signing Your Android Applications

All Android applications must be digitally signed before they are allowed to be deployed onto a device (or emulator). Unlike some mobile platforms, you need not purchase digital certificates from a certificate authority (CA) to sign your applications. Instead, you can generate your own self-signed certificate and use it to sign your Android applications.

When you use Eclipse to develop your Android application and then press F11 to deploy it to an emulator, Eclipse automatically signs it for you. You can verify this by going to Windows ⇨ Preferences in Eclipse, expanding the Android item, and selecting Build (see Figure 6-1). Eclipse uses a default debug keystore (appropriately named "`debug.keystore`") to sign your application. A keystore is commonly known as a *digital certificate*.

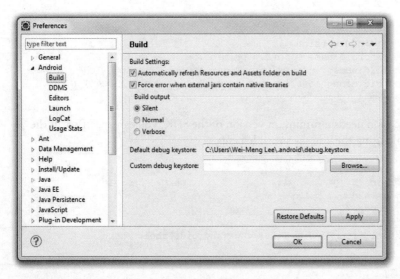

FIGURE 6-1

If you are publishing an Android application, you must sign it with your own certificate. Applications signed with the debug certificate cannot be published. While you can manually generate your own certificates using the `keytool.exe` utility provided by the Java SDK, Eclipse has made it easy for you by including a wizard that walks you through the steps to generate a certificate. It will also sign your application with the generated certificate (which you can also sign manually using the `jarsigner.exe` tool from the Java SDK).

The following Try It Out demonstrates how to use Eclipse to export an Android application and sign it with a newly generated certificate.

TRY IT OUT   **Exporting and Signing an Android Application**

1. Using Eclipse, open the LBS projected created in Chapter 4.

2. Select the LBS project in Eclipse and then select File ⇨ Export. . . .

3. In the Export dialog, expand the Android item and select Export Android Application (see Figure 6-2). Click Next.

4. The LBS project should now be displayed (see Figure 6-3). Click Next.

FIGURE 6-2

FIGURE 6-3

5. Select the "Create new keystore" option to create a new certificate (keystore) for signing your application (see Figure 6-4). Enter a path to save your new keystore and then enter a password to protect the keystore. For this example, enter **password** as the password. Click Next.

6. Provide an alias for the private key (name it **DistributionKeyStoreAlias**; see Figure 6-5) and enter a password to protect the private key. For this example, enter **password** as the password. You also need to enter a validity period for the key. According to Google, your application must be signed with a cryptographic private key whose validity period ends after 22 October 2033. Hence, enter a number that is greater than 2033 minus the current year. Click Next.

FIGURE 6-4

7. Enter a path to store the destination APK file (see Figure 6-6). Click Finish. The APK file will now be generated.

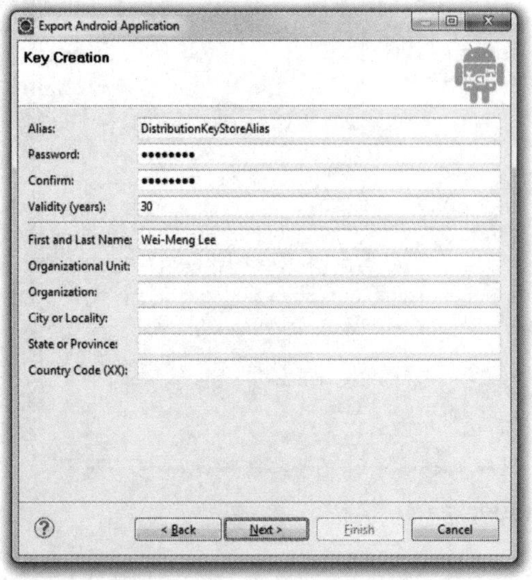

FIGURE 6-5

FIGURE 6-6

8. Recall from Chapter 4 that the LBS application requires the use of the Google Maps API key, which you applied by using your debug.keystore's MD5 fingerprint. This means that the Google Maps API key is essentially tied to the debug.keystore used to sign your application. Because you are now generating your new keystore to sign your application for deployment, you need to apply for the Google Maps API key again, using the new keystore's MD5 fingerprint. To do so, go to the command prompt and enter the following command (the location of your keytool.exe utility might differ slightly, in which case you would need to replace the path of the keystore with the path you selected earlier in step 5; see also Figure 6-7):

```
C:\Program Files\Java\jre6\bin>keytool.exe -list -alias DistributionKeyStoreAlias
-keystore "C:\Users\Wei-Meng Lee\Desktop\DistributionKeyStore" -storepass
password -keypass password
```

FIGURE 6-7

9. Using the MD5 fingerprint obtained from the previous step, go to http://code.google.com/android/add-ons/google-apis/maps-api-signup.html and sign up for a new Maps API key.

10. Enter the new Maps API key in the main.xml file:

```xml
<?xml version="1.0" encoding="utf-8"?>
<LinearLayout xmlns:android="http://schemas.android.com/apk/res/android"
    android:orientation="vertical"
    android:layout_width="fill_parent"
    android:layout_height="fill_parent"
    >
<com.google.android.maps.MapView
    android:id="@+id/mapView"
    android:layout_width="fill_parent"
    android:layout_height="fill_parent"
    android:enabled="true"
    android:clickable="true"
    android:apiKey="<Your Key Here>" />
</LinearLayout>
```

11. With the new Maps API key entered in the `main.xml` file, you now need to export the application once more and resign it. Repeat steps 2 through 4. When you are asked to select a keystore, select the "Use existing keystore" option (see Figure 6-8) and enter the password you used earlier to protect your keystore (in this case, "`password`"). Click Next.

12. Select the "Use existing key" option (see Figure 6-9) and enter the password you set earlier to secure the private key (enter "`password`"). Click Next.

13. Click Finish (see Figure 6-10) to generate the APK file again.

FIGURE 6-8

FIGURE 6-9

FIGURE 6-10

That's it! The APK is now generated and it contains the new Map API key that is tied to the new keystore.

How It Works

Eclipse provides the Export Android Application option, which helps you to both export your Android application as an APK file and generate a new keystore to sign the APK file. For applications that use the Maps API key, note that the Maps API key must be associated with the new keystore that you use to sign your APK file.

DEPLOYING APK FILES

Once you have signed your APK files, you need a way to get them onto your users' devices. The following sections describe the various ways to deploy your APK files. Three methods are covered:

➤ Deploying manually using the `adb.exe` tool

➤ Hosting the application on a web server

➤ Publishing through the Android Market

Besides these methods, you can install your applications on users' devices through e-mails, SD card, and so on. As long as you can transfer the APK file onto the user's device, you can install the application.

Using the adb.exe Tool

Once your Android application is signed, you can deploy it to emulators and devices using the `adb.exe` (Android Debug Bridge) tool (located in the `platform-tools` folder of the Android SDK).

Using the command prompt in Windows, navigate to the *<Android_SDK>\platform-tools* folder. To install the application to an emulator/device (assuming the emulator is currently up and running or a device is currently connected), issue the following command:

```
adb install "C:\Users\Wei-Meng Lee\Desktop\LBS.apk"
```

EXPLORING THE ADB.EXE TOOL

The `adb.exe` tool is a very versatile tool that enables you to control Android devices (and emulators) connected to your computer.

By default, when you use the `adb` command, it assumes that currently there is only one connected device/emulator. If you have more than one device connected, the `adb` command returns an error message:

```
error: more than one device and emulator
```

You can view the devices currently connected to your computer by using the `devices` option with `adb`, like this:

```
D:\Android 3.0\android-sdk-windows\platform-tools>adb devices
List of devices attached
HT07YPY09335    device
emulator-5554   device
emulator-5556   device
```

continues

(continued)

As the preceding example shows, this returns the list of devices currently attached. To issue a command for a particular device, you need to indicate the device using the -s option, like this:

```
adb -s emulator-5556 install LBS.apk
```

If you try to install an APK file onto a device that already has the APK file, it will display the following error message:

```
Failure [INSTALL_FAILED_ALREADY_EXISTS]
```

When you inspect the Launcher on the Android device/emulator, you will be able to see the LBS icon (on the top of Figure 6-11). If you select Settings ⇨ Applications ⇨ Manage Applications on your Android device/emulator, you will see the "Where Am I" application (on the bottom of Figure 6-11).

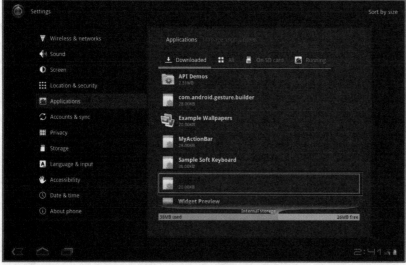

FIGURE 6-11

Besides using the `adb.exe` tool to install applications, you can also use it to remove an installed application. To do so, use the `shell` option to remove an application from its installed folder:

```
adb shell rm /data/app/net.learn2develop.LBS.apk
```

Another way to deploy an application is to use the DDMS tool in Eclipse (see Figure 6-12). With an emulator (or device) selected, use the File Explorer in DDMS to go to the `/data/app` folder and use the "Push a file onto the device" button to copy the APK file onto the device.

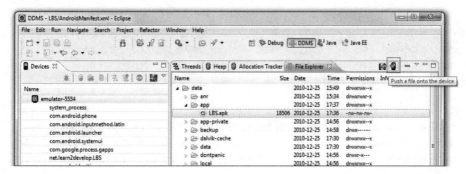

FIGURE 6-12

Using a Web Server

If you wish to host your application on your own, you can use a web server to do that. This is ideal if you have your own web hosting services and want to provide the application free of charge to your users (or you can restrict access to certain groups of people).

 NOTE *Even if you restrict your application to a certain group of people, there is nothing to stop users from redistributing your application to other users after they have downloaded your APK file.*

To demonstrate this, I will use the Internet Information Server (IIS) on my Windows 7 computer. Copy the signed `LBS.apk` file to `c:\inetpub\wwwroot\`. In addition, create a new HTML file named `Install.html` with the following content:

```html
<html>
<title>Where Am I application</title>
<body>
Download the Where Am I application <a href="LBS.apk">here</a>
</body>
</html>
```

NOTE *If you are unsure how to set up the IIS on your Windows 7 computer, check out the following link:* `http://technet.microsoft.com/en-us/library/cc725762.aspx`.

On your web server, you may need to register a new MIME type for the APK file. The MIME type for the .apk extension is `application/vnd.android.package-archive`.

 NOTE *To install APK files over the Web, you need an SD card installed on your emulator or device. This is because the downloaded APK files will be saved to the* `download` *folder created on the SD card.*

By default, for online installation of Android applications, the Android emulator or device only allows applications to be installed from the Android Market (`www.android.com/market/`). Hence, for installation over a web server, you need to configure your Android emulator/device to accept applications from non-Market sources.

From the Application settings menu, check the "Unknown sources" item (see Figure 6-13). You will be prompted with a warning message. Click OK. Checking this item will allow the emulator/device to install applications from other non-Market sources (such as from a web server).

FIGURE 6-13

To install the LBS.apk application from the IIS web server running on your computer, launch the Browser application on the Android emulator/device and navigate to the URL pointing to the APK file. To refer to the computer running the emulator, you should use the computer's IP address. Figure 6-14 shows the Install.html file loaded on the web browser. Clicking the "here" link will download the APK file onto your device. Click the Download button at the bottom of the screen to reveal the download's status.

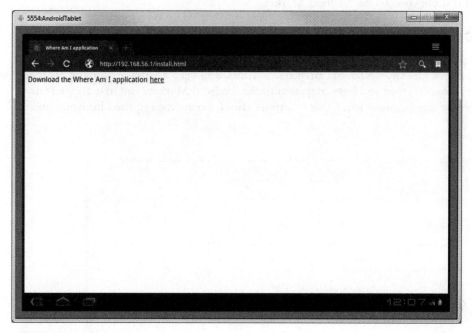

FIGURE 6-14

To install the downloaded application, simply tap on it. It will show the permission(s) required by the application. Click the Install button to proceed with the installation. When the application is installed, you can launch it by clicking the Open button.

Besides using a web server, you can also e-mail your application to users as an attachment; when the users receive the e-mail, they can download the attachment and install the application directly onto their device.

Publishing on the Android Market

So far, you have learned how to package your Android application and distribute it in various ways — via web server, the adb.exe file, e-mail, and SD card.

However, these methods do not provide a way for users to discover your applications easily. A better way is to host your application on the Android Market, a Google-hosted service that makes it very easy for users to discover and download (i.e., purchase) applications for their Android devices. Users simply need to launch the Market application on their Android device in order to discover a wide range of applications that they can install on their devices.

In this section, you will learn how to publish your Android application on the Android Market. You will walk through each of the steps involved, including the various items you need to prepare your application for submission to the Android Market.

Creating a Developer Profile

The first step toward publishing on the Android Market is to create a developer profile at `http://market.android.com/publish/Home`. For this, you need a Google account (such as your Gmail account). Once you have logged in to the Android Market, you first create your developer profile (see Figure 6-15). Click Continue after entering the required information.

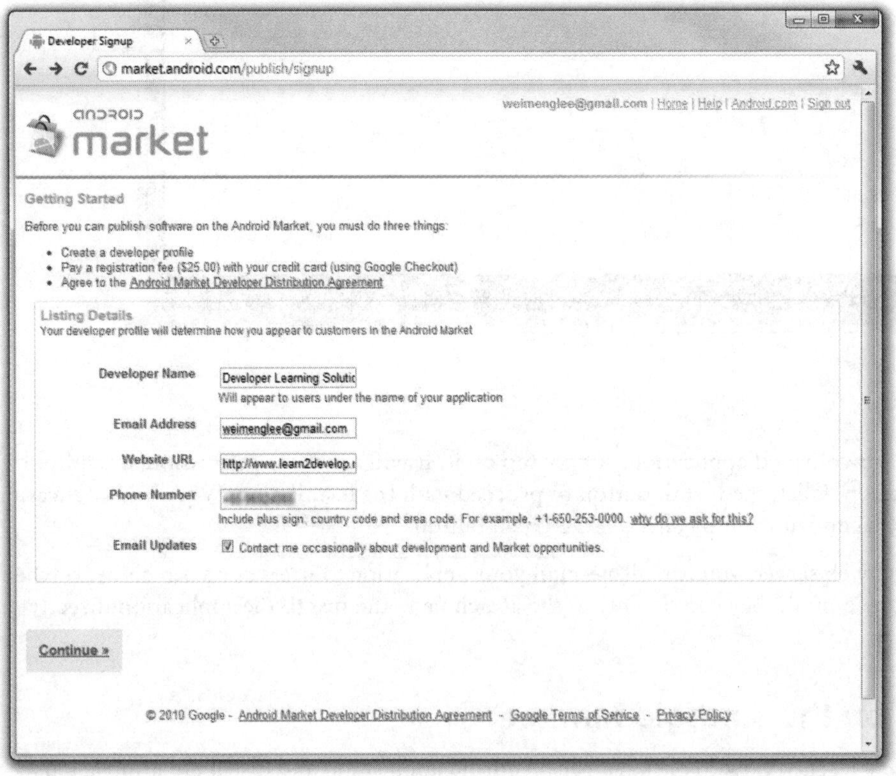

FIGURE 6-15

For publishing on the Android Market, you need to pay a one-time registration fee, currently U.S. $25. Click the Google Checkout button to be redirected to a page where you can pay the registration fee. After paying, click the Continue link.

Next, you need to agree to the Android Market Developer Distribution Agreement. Check the "I agree" checkbox and then click the "I agree. Continue" link.

Submitting Your Apps

Once you have set up your profile, you are ready to submit your application to the Android Market. If you intend to charge for your application, click the Setup Merchant Account link located at the bottom of the screen. Here you enter additional information such as bank account and tax ID.

For free applications, click the Upload Application link, shown in Figure 6-16.

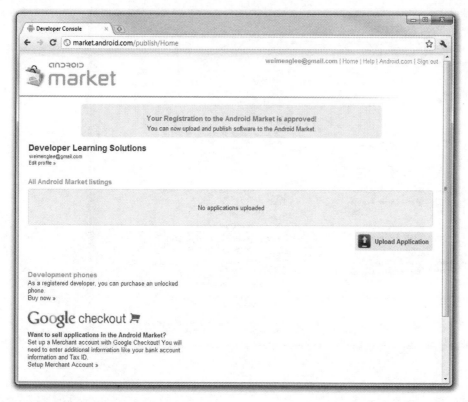

FIGURE 6-16

You will be asked to supply some details about your application. Figure 6-17 shows the first set of details you need to provide. Among the information needed, the following are compulsory:

➤ The application in APK format.

➤ At least two screenshots. You can use the DDMS perspective in Eclipse to capture screenshots of your application running on the emulator or real device.

➤ A high-resolution application icon. This size of this image must be 512 × 512 pixels.

The other information details are optional, and you can always supply them later.

FIGURE 6-17

Figure 6-18 shows that I have uploaded the LBS.apk file to the Android Market site. In particular, note that based on the APK file that you have uploaded, users are warned about any specific permissions required, and your application's features will be used to filter search results. For example, because my application requires GPS access, it will not appear in the search result list if a user searches for my application on a device that does not have a GPS receiver.

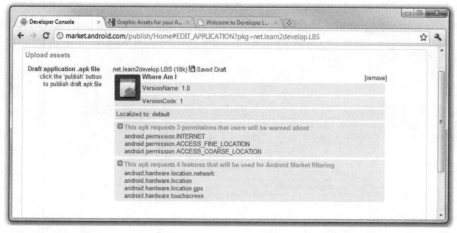

FIGURE 6-18

The next set of information you need to supply, shown in Figure 6-19, includes the title of your application, its description, as well as recent changes' details (useful for application updates). You can also select the application type and the category in which it will appear in the Android Market.

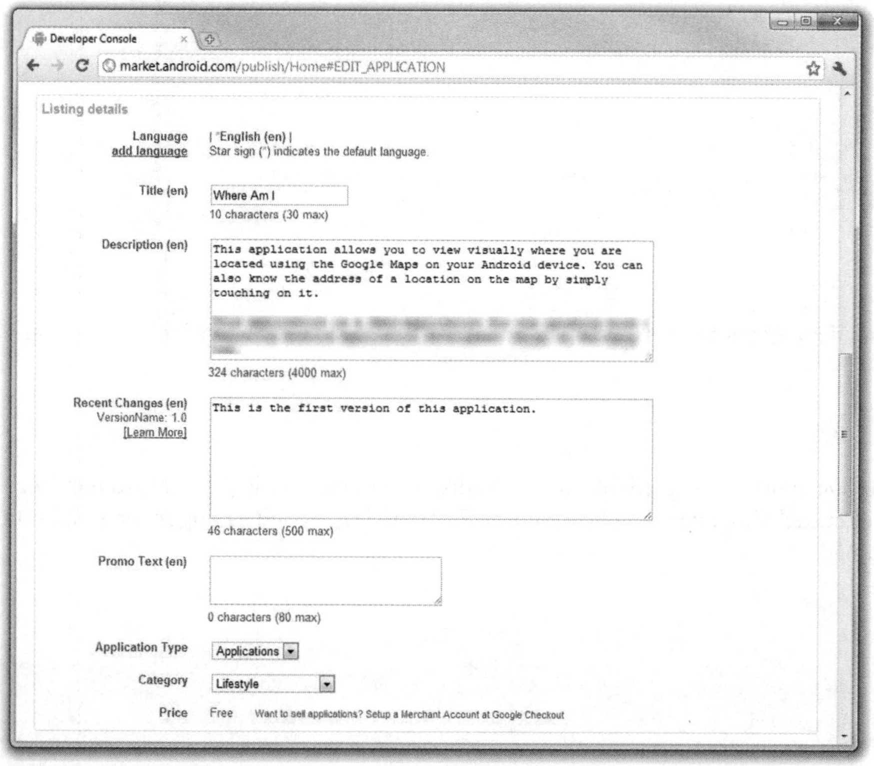

FIGURE 6-19

On the last dialog, you indicate whether your application employs copy protection, and specify a content rating. You also supply your website URL and your contact information (see Figure 6-20). When you have given your consent to the two guidelines and agreements, click Publish to publish your application on the Android Market.

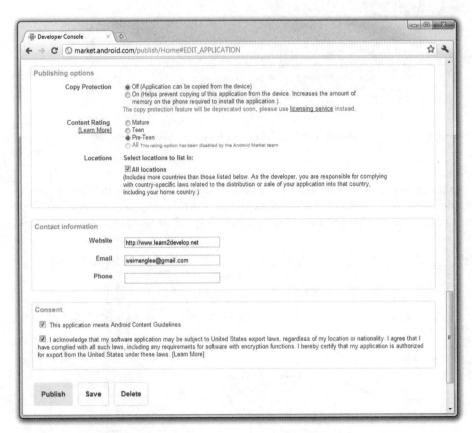

FIGURE 6-20

That's it! Your application is now available on the Android Market. You will be able to monitor any comments submitted about your application (see Figure 6-21), as well as bug reports and total number of downloads.

FIGURE 6-21

Good luck! All you need to do now is wait for the good news; and hopefully you can laugh your way to the bank soon!

SUMMARY

In this chapter, you have learned how you can export your Android application as an APK file and then digitally sign it with a keystore you create yourself. You then learned about the various ways you can distribute your application, and the advantages of each method. Finally, you walked through the steps required to publish on the Android Market, which makes it possible for you to sell your application and reach out to a wider audience. Hopefully, this exposure enables you to sell a lot of copies and thereby make some decent money!

EXERCISES

1. How do you specify the minimum version of Android required by your application?

2. How do you generate a self-signed certificate for signing your Android application?

3. How do you configure your Android device to accept applications from non-Market sources?

Answers to the Exercises can be found in Appendix C.

▶ WHAT YOU LEARNED IN THIS CHAPTER

TOPIC	KEY CONCEPTS
Checklist for publishing your apps	To publish an application on the Android Market, an application must have the four required attributes in the `AndroidManifest.xml` file: ➤ `android:versionCode` ➤ `android:versionName` ➤ `android:icon` ➤ `android:label`
Signing applications	All applications to be distributed must be signed with a self-signed certificate. The debug keystore is not valid for distribution.
Exporting an application and signing it	Use the Export feature of Eclipse to export the application as an APK file and then sign it with a self-signed certificate.
Deploying APK files	You can deploy using various means, including web server, e-mail, `adb.exe`, DDMS, etc.
Publishing your application on the Android Market	Apply for the Android Market with a one-time fee of U.S.$25 and you will be able to sell and host your apps on the Android Market.

PART III
Appendices

Using Eclipse for Android Development

Although Google supports the development of Android applications using IDEs such as IntelliJ, or basic editors like Emacs, Google's recommendation is to use the Eclipse IDE together with the Android Development Tools Plugin. Doing so makes developing Android applications much easier and more productive. This appendix describes some of the neat features available in Eclipse that can make your development life much easier.

 NOTE *If you have not downloaded Eclipse yet, please start with Chapter 1, where you will learn how to obtain Eclipse and configure it to work with the Android SDK. This appendix assumes that you have already set up your Eclipse environment for Android development.*

GETTING AROUND IN ECLIPSE

Eclipse is a highly extensible multi-language software development environment that supports application development of all sorts. Using Eclipse, you could write and test your applications using a wide variety of languages, such as Java, C, C++, PHP, Ruby, and so on. Because of its extensibility, new users of Eclipse often feel inundated with the IDE. Hence, the following sections aim to make you more at home with Eclipse when you develop your Android applications.

Workspaces

Eclipse adopts the concept of a *workspace*. A workspace is a folder that you have chosen to store all your projects.

When you first start Eclipse, you are prompted to select a workspace (see Figure A-1).

When Eclipse has finished launching the projects located in your workspace, it will display several panes in the IDE (see Figure A-2).

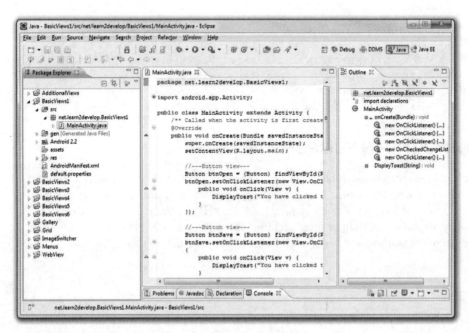

FIGURE A-2

The following sections highlight some of the more important panes that you need to know about when developing Android applications.

Package Explorer

The Package Explorer, shown in Figure A-3, lists all the projects currently in your workspace. To edit a particular item in your project, you can double-click on it and the file will be displayed in the respective editor.

You can also right-click on each item displayed in the Package Explorer to display context sensitive menu(s) related to the selected item. For example, if you wish to add a new .java file to the project, you can right-click on the package name in the Package Explorer and then select New ⇨ Class (see Figure A-4).

FIGURE A-3

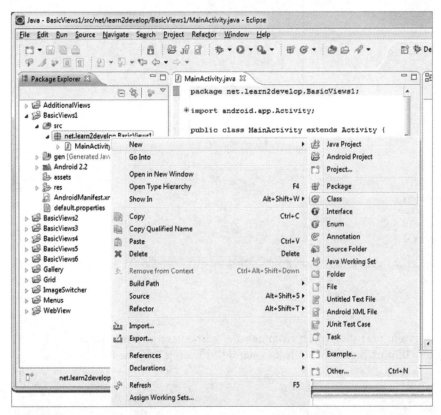

FIGURE A-4

Using Projects from Other Workspaces

There may be times when you have several workspaces created to store different projects. If you need to access the project in another workspace, there are generally two ways to go about doing so. First, you can switch to the desired workspace by selecting File ⇨ Switch Workspace (see Figure A-5). Specify the new workspace to work on and then restart Eclipse.

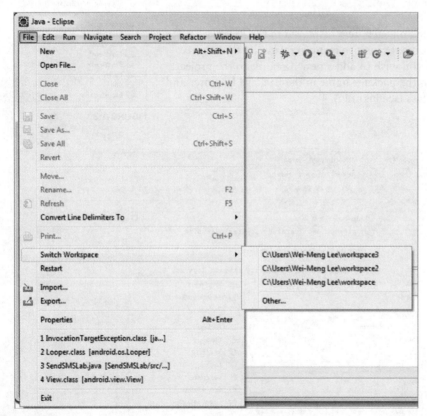

FIGURE A-5

The second method is to import the project from another workspace into the current one. To do so, select File ⇨ Import... and then select General ⇨ Existing Projects into Workspace (see Figure A-6).

In the Select root directory textbox, enter the path of the workspace containing the project(s) you want to import and tick the project(s) you want to import (see Figure A-7). To import the selected project(s), click Finish.

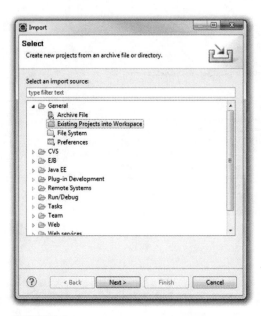

FIGURE A-6

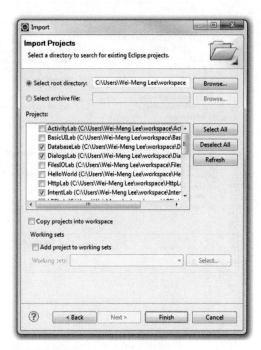

FIGURE A-7

Note that even when you import a project from another workspace into the current workspace, the physical location of the imported project remains unchanged. That is, it will still be located in its original directory. To have a copy of the project in the current workspace, check the "Copy projects into workspace" option.

Editors

Depending on the type of items you have double-clicked in the Package Explorer, Eclipse will open the appropriate editor for you to edit the file. For example, if you double-click on a .java file, the text editor for editing the source file will be opened (see Figure A-8).

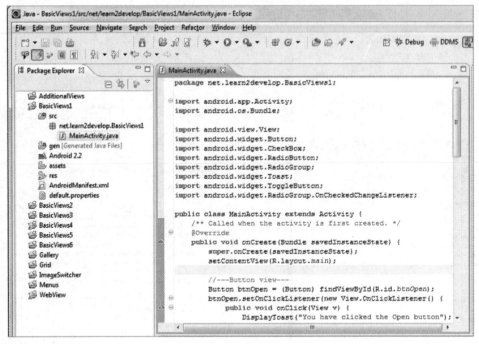

FIGURE A-8

If you double-click on the `icon.png` file in the `res/drawable-mdpi` folder, the Windows Photo Viewer application will be invoked to display the image (see Figure A-9).

If you double-click on the `main.xml` file in the `res/layout` folder, Eclipse will display the UI editor, where you can graphically view and build the layout of your UI (see Figure A-10).

FIGURE A-9

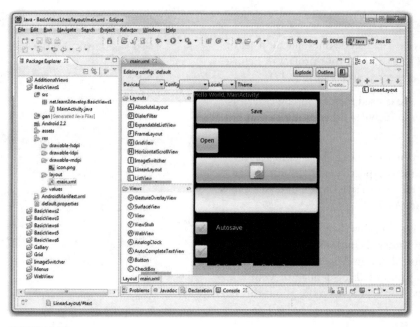

FIGURE A-10

To edit the UI manually using XML, you can switch to XML view by clicking on the main.xml tab located at the bottom of the screen (see Figure A-11).

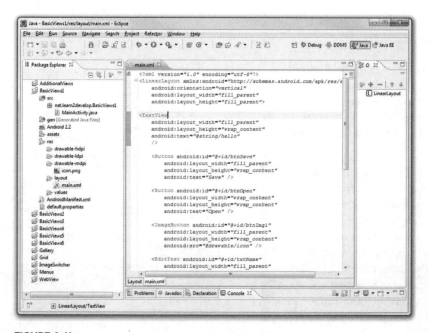

FIGURE A-11

Perspectives

In Eclipse, a *perspective* is a visual container for a set of views and editors. When you edit your project in Eclipse, you are in the Java perspective (see Figure A-12).

FIGURE A-12

The Java EE perspective is used for developing enterprise Java applications, and it includes other modules that are relevant to it.

You can switch to other perspectives by clicking on the perspective name. If the perspective name is not shown, you can click the Open Perspective button and add a new perspective (see Figure A-13).

The DDMS perspective contains the tools for communicating with Android emulators and devices. This is covered in more detail in Appendix B. The Debug perspective contains panes used for debugging your Android applications. You will learn more about this later in this appendix.

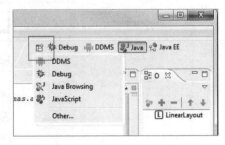

FIGURE A-13

Auto Import of Namespaces

The various classes in the Android library are organized into namespaces. As such, when you use a particular class from a namespace, you need to import the appropriate namespaces, like this:

```
import android.app.Activity;

import android.os.Bundle;
```

Because the number of classes in the Android Library is very large, remembering the correct namespace for each class is not an easy task. Fortunately, Eclipse can help find the correct namespace for you, enabling you to import it with just a click.

Figure A-14 shows that I have declared an object of type `Button`. Because I did not import the correct namespace for the `Button` class, Eclipse signals an error beneath the statement. When you move the mouse over the `Button` class, Eclipse displays a list of suggested fixes. In this case, I need to import the `android.widget.Button` namespace. Clicking the "Import 'Button' (android .widget)" link will add the import statement at the top of the file.

FIGURE A-14

Alternatively, you can also use the following key combination: Control+Shift+o. This key combination will cause Eclipse to automatically import all the namespaces required by your class.

Code Completion

Another very useful feature of Eclipse is the support for code completion. Code completion displays a context-sensitive list of relevant classes, objects, methods, and property names as you type in the code editor. For example, Figure A-15 shows code-completion in action. As I type the word "`fin`," I can activate the code-completion feature by pressing Ctrl+Space. This will bring up a list of names that begin with "`fin`."

To select the required name, simply double-click on it or use your cursor to highlight it and press the Enter key.

Code completion also works when you type a "." after an object/class name. Figure A-16 shows an example.

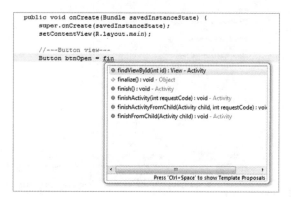

FIGURE A-15 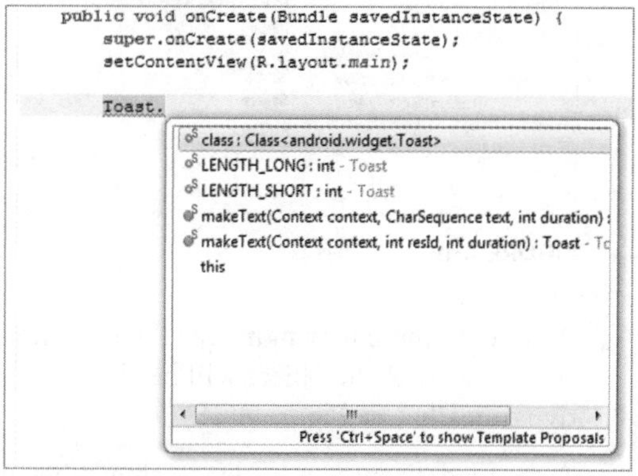   **FIGURE A-16**

Refactoring

Refactoring is a very useful feature that most modern IDEs support. Eclipse supports a whole slew of refactoring features that make application development efficient.

In Eclipse, when you position the cursor at a particular object/variable, the editor will highlight all occurrences of the selected object in the current source (see Figure A-17).

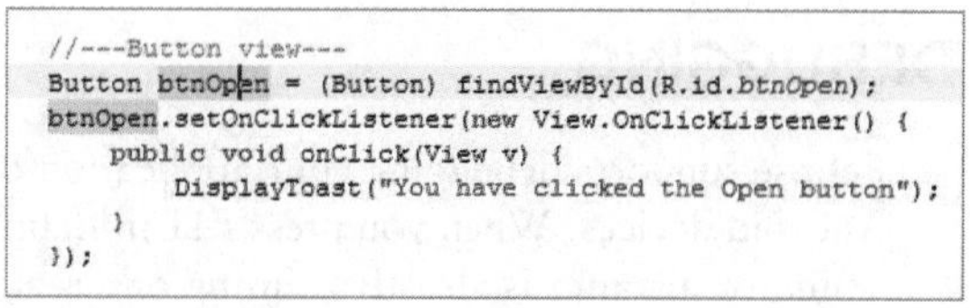

```
//---Button view---
Button btnOpen = (Button) findViewById(R.id.btnOpen);
btnOpen.setOnClickListener(new View.OnClickListener() {
    public void onClick(View v) {
        DisplayToast("You have clicked the Open button");
    }
});
```

FIGURE A-17

This feature is very useful for identifying where a particular object is used in your code. To change the name of an object, simply right-click on it and select Refactor ➪ Rename. . . (see Figure A-18).

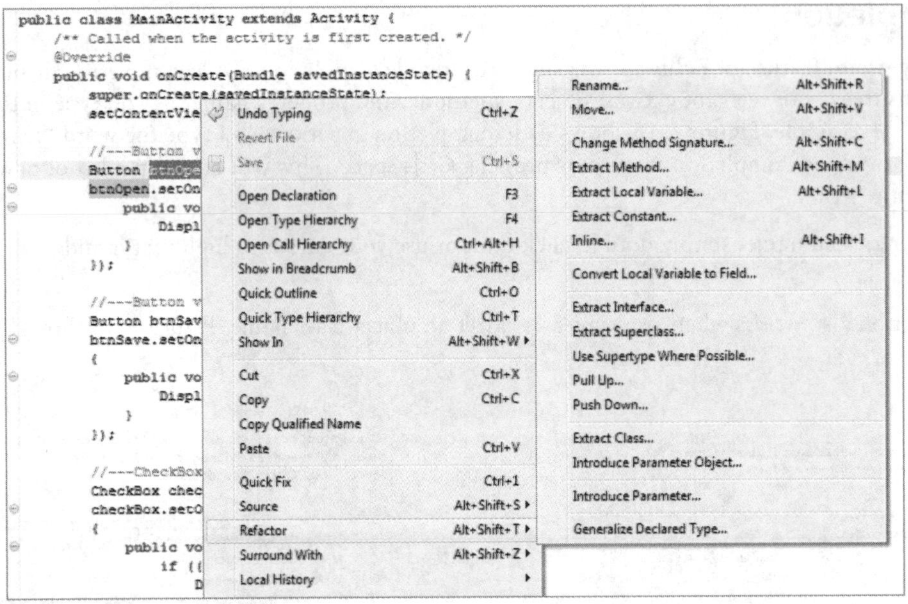

FIGURE A-18

After entering a new name for the object, all occurrences of the object will be changed dynamically (see Figure A-19).

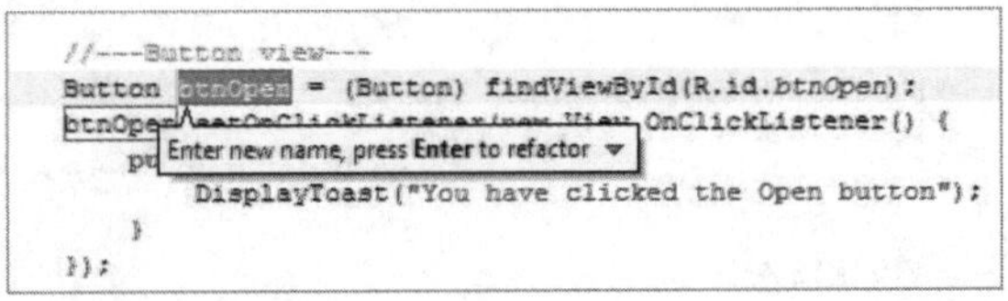

A detailed discussion of refactoring is beyond the scope of this book. For more information on refactoring in Eclipse, refer to www.ibm .com/developerworks/library/os-ecref/.

FIGURE A-19

DEBUGGING

Eclipse supports debugging your application on both the Android emulators as well as on real Android devices. When you press F11 in Eclipse, Eclipse will first determine whether an Android emulator instance is already running or a real device is connected. If at least one emulator (or device) is running, Eclipse will deploy the application onto the running emulator or the connected device. If there is no emulator running and no connected device, Eclipse will automatically launch an instance of the Android emulator and deploy the application onto it.

If you have more than one emulator or device connected, Eclipse will prompt you to select the target emulator/device on which to deploy the application (see Figure A-20). Select the target device you want to use and click OK.

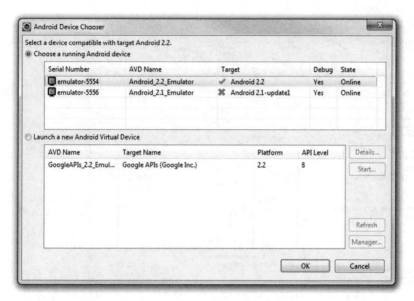

FIGURE A-20

If you want to launch a new emulator instance to test the application, select Window ⇨ Android SDK and AVD Manager to launch the AVD manager.

Setting Breakpoints

Setting breakpoints is a good way to temporarily pause the execution of the application and then examine the content of variables and objects.

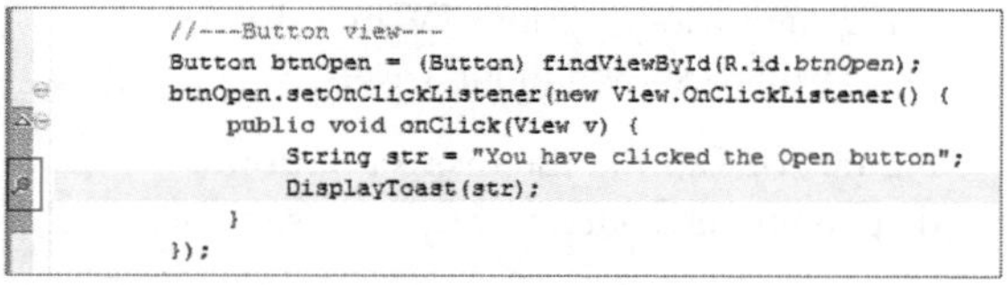

```
//---Button view---
Button btnOpen = (Button) findViewById(R.id.btnOpen);
btnOpen.setOnClickListener(new View.OnClickListener() {
    public void onClick(View v) {
        String str = "You have clicked the Open button";
        DisplayToast(str);
    }
});
```

FIGURE A-21

To set a breakpoint, double-click on the leftmost column in the code editor. Figure A-21 shows a breakpoint set on a particular statement.

When the application is running and the first breakpoint is reached, Eclipse will display a Confirm Perspective Switch dialog. Basically, it wants to switch to the Debug perspective. To prevent this window from appearing again, check the "Remember my decision" checkbox at the bottom and click Yes.

Eclipse now highlights the breakpoint (see Figure A-22).

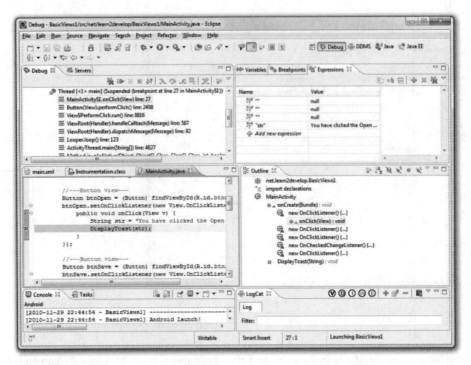

FIGURE A-22

At this point, you can right-click on any selected object/variable and view its content using the various options (Watch, Inspect, and Display) shown in Figure A-23.

Figure A-24 shows the Inspect option displaying the content of the str variable.

There are several options at this point to continue the execution:

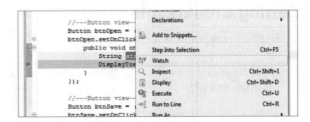

FIGURE A-23

➤ **Step Into** — Press F5 to step into the next method call/statement.

➤ **Step Over** — Press F6 to step over the next method call without entering it.

➤ **Step Return** — Press F7 to return from a method that has been stepped into.

➤ **Resume Execution** — Press F8 to resume the execution.

```
//---Button view---
Button btnOpen = (Button) findViewById(R.id.btnOpen);
btnOpen.setOnClickListener(new View.OnClickListener() {
    public void onClick(View v) {
        String str = "You have clicked the Open button";
        DisplayToast(str);
    }
});

//---Button view---
Button btnSave = (Bu
btnSave.setOnClickLi
{
    public void onCl
        DisplayToast
    }
});

//---CheckBox---
```

▲ ○ str= "You have clicked the Open button" (id=830067779976)
   ⬛ count= 32
   ⬛ hashCode= 561881161
   ⬛ offset= 0
 ▷ ⬛ value= (id=830067780008)

You have clicked the Open button

FIGURE A-24

Exceptions

As you develop in Android, you will encounter numerous run-time exceptions that prevent your program from continuing. Examples of run-time exceptions include the following:

➤ Null reference exception (accessing an object which is null)

➤ Failure to specify the required permissions required by your application

➤ Arithmetic operation exceptions

Figure A-25 shows the current state of an application when an exception occurred. In this example, I am trying to send an SMS message from my application and it crashes when the SMS message is about to be sent.

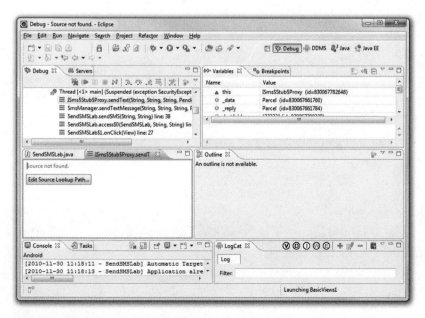

FIGURE A-25

The various windows do not really identify the cause of the exception. To find out more, press F6 in Eclipse so that it can step over the current statement. The Variables window, shown in Figure A-26, indicates the cause of the exception. In this case, the SEND_SMS permission is missing.

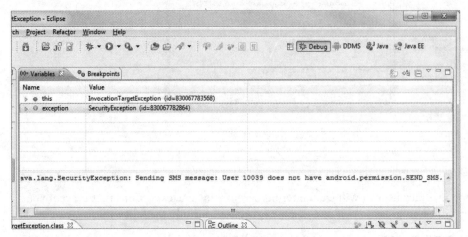

FIGURE A-26

To remedy this, all you need to do is to add the following permission statement in the AndroidManifest.xml file:

```
<uses-permission
    android:name="android.permission.SEND_SMS"/>
```

B
Using the Android Emulator

The Android emulator ships with the Android SDK and is an invaluable tool to help test your application without requiring you to purchase a real device. While you should thoroughly test your applications on real devices before you deploy them, the emulator mimics most of the capabilities of real devices. It is a very handy tool that you should make use of during the development stage of your project. This appendix provides some common tips and tricks for mastering the Android emulator.

USES OF THE ANDROID EMULATOR

As discussed in Chapter 1, you can use the Android emulator to emulate the different Android configurations by creating Android Virtual Devices (AVDs).

You launch the Android emulator by directly starting the AVD you have created in the Android SDK and AVD Manager window (see Figure B-1). Simply select the AVD and click the Start button. You have the option to scale the emulator to a particular size and monitor DPI.

Alternatively, when you run an Android project in Eclipse, the Android emulator is automatically invoked to test your application. You can customize the Android emulator for each of your Android projects in Eclipse. To do so, simply select Run ⇨ Run Configurations. Select the project name listed under Android Application on the left (see Figure B-2), and on the right you will see the Target tab. You can choose your preferred AVD to use for testing your application, as well as emulate different scenarios such as network speed and network latency.

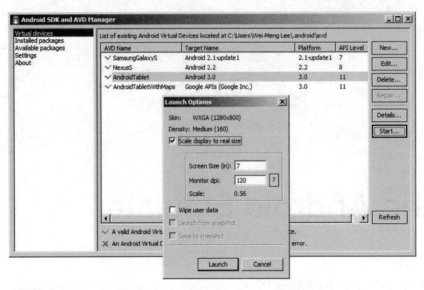

FIGURE B-1

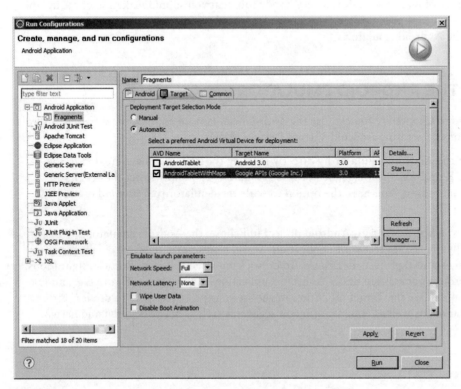

FIGURE B-2

CREATING SNAPSHOTS

In the latest version of the AVD Manager, you now have the option to save an emulator's state to a snapshot file. Saving an emulator's state to a snapshot file enables the emulator to be started quickly the next time you try to launch it, effectively bypassing the lengthy boot-up time. This is especially useful for the Android 3.0 emulator, which can take up to five minutes to boot up.

To use the snapshot feature, simply check the Snapshot Enabled checkbox when you create a new AVD (see Figure B-3).

When you launch the AVD from the Start . . . button, check the "Launch from snapshot" and "Save to snapshot" checkboxes (see Figure B-4). The first time you launch the emulator, it will boot up normally. When you close the emulator, it will now save the state to a snapshot file. The next time you launch the emulator, it will appear almost instantly, restoring its state from the snapshot file.

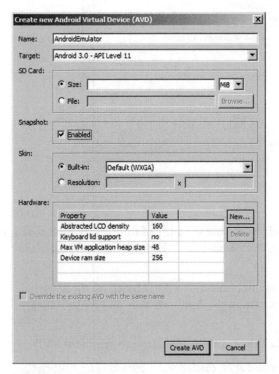

FIGURE B-3

FIGURE B-4

INSTALLING CUSTOM AVDS

Sometimes device manufacturers provide their own AVDs that you can use to emulate your applications running on their devices. A good example is Samsung, which provides the Samsung Galaxy Tab add-on for emulating their Samsung Galaxy Tab tablet. To install the Samsung Galaxy Tab add-on, first launch the Android SDK and AVD Manager in Eclipse, and then select the Available Packages item on the left side of the dialog (see Figure B-5).

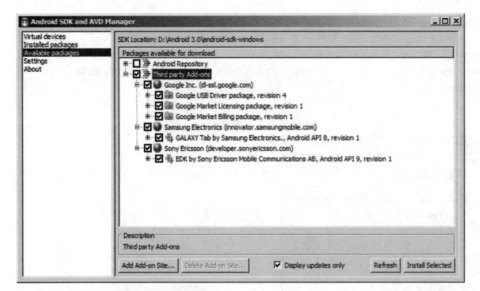

FIGURE B-5

Check the "Third party Add-ons" checkbox and you should see a list of third-party tools that you can download for testing. At the time of writing, both Samsung and Sony Ericsson provide their own AVD for testing your apps on their devices.

After the downloaded packages are installed, you can create a new AVD based on the newly downloaded package. Select the Virtual Devices item in the Androids SDK and AVD Manager window and click the New . . . button.

Name the new AVD as shown in Figure B-6. Click the Create AVD button to create the AVD.

To launch the SamsungGalaxyTab AVD, select it and click the Start . . . button. The Launch Options dialog will appear. Check the "Scale display to real size" option if you want to resize the emulator. This is very useful if you are running the emulator on a small monitor (such as a notebook computer). Specify a screen size and click the Launch button to start the emulator. Figure B-7 shows the Samsung Galaxy Tab emulator.

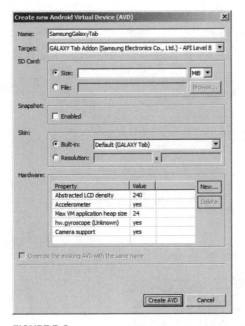

FIGURE B-6

FIGURE B-7

EMULATING REAL DEVICES

Besides using the Android emulator to test the different configurations of Android, you can also make use of the emulator to emulate real devices, using the system images provided by device manufacturers.

For example, HTC provides images for their devices running Android 1.5 and 1.6 (http://developer.htc.com/google-io-device.html#s3). You can download a device's system image and then use the Android emulator to emulate it using the system image. Here is how this can be done (in theory, this should work for any version of Android).

 NOTE *If you use HTC's image, you should be able to boot up the emulator without problems. However, the network cannot be enabled. Some kind souls have uploaded a modified image that works properly. You can try downloading it at* www.4shared.com/get/x6pZm3-W/system.html.

First, using the Android SDK and AVD Manager, create a new AVD. In the case of HTC, create an AVD using Android 1.6 as the platform. The AVD will be located in the

C:\Users\<*username*>\.android\avd\<*avd_name*>.avd folder. As shown in Figure B-8, a newly created AVD contains only two files in the folder.

Using the downloaded system image, copy the system.img file into the AVD folder (see Figure B-9).

FIGURE B-8

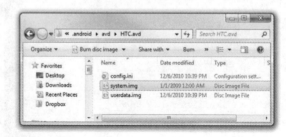

FIGURE B-9

Launch the AVD and you should see it booting up (see Figure B-10).

FIGURE B-10

You can proceed to sign in using your Google account. When prompted to slide open the keyboard, press Ctrl+F11 to change the orientation of the emulator. This action tricks the emulator into believing that you are sliding the keyboard open. Once you have successfully signed in, you will be able to explore the Android Market on your emulator (see Figure B-11)!

FIGURE B-11

SD CARD EMULATION

When you create a new AVD, you can emulate the existence of an SD card (see Figure B-12). Simply enter the size of the SD card that you want to emulate (in the figure, it is 200MB).

Alternatively, you can simulate the presence of an SD card in the Android emulator by creating a disk image first and then attaching it to the AVD. The `mksdcard` `.exe` utility (also located in the `tools` folder of the Android SDK) enables you to create an ISO disk image. The following command creates an ISO image that is 2GB in size (see also Figure B-13):

```
mksdcard 2048M sdcard.iso
```

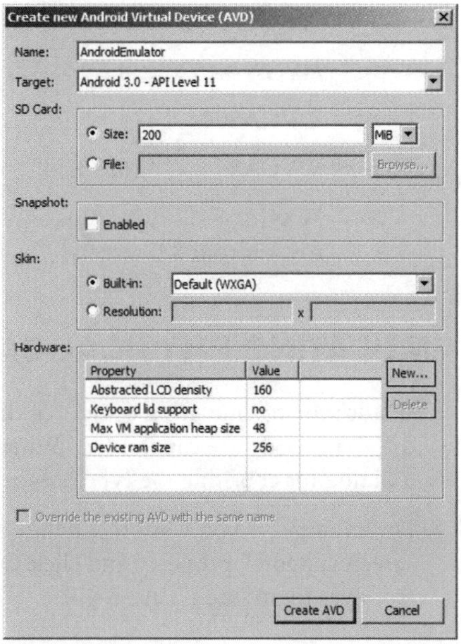

FIGURE B-12

FIGURE B-13

Once the image is created, you can specify the location of the ISO file, as shown in Figure B-14.

EMULATING DEVICES WITH DIFFERENT SCREEN SIZES

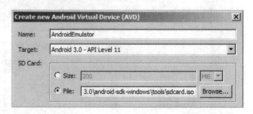

FIGURE B-14

Besides emulating an SD card, you can also emulate devices with different screen sizes. Figure B-15 indicates that the AVD is emulating the WXGA skin, which has a resolution of 1280 × 800 pixels. Note that the LCD density is 160, which means that this screen has a pixel density of 160 pixels per inch.

For each target that you select, a list of skins is available. The Android SDK supports the following screen resolutions:

- ➤ **QVGA** — 240 × 320
- ➤ **WQVGA400** — 240 × 400
- ➤ **WQVGA432** — 240 × 432
- ➤ **HVGA** — 320 × 480
- ➤ **WVGA800** — 480 × 800
- ➤ **WVGA854** — 480 × 854
- ➤ **WXGA** — 1280 × 800 (only applicable for Android 3.0 targets)

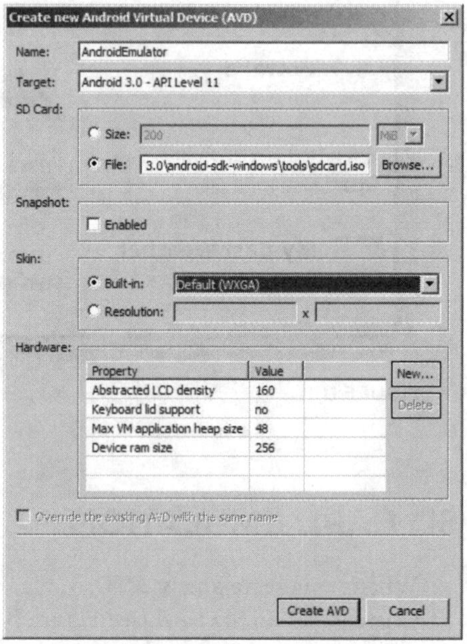

FIGURE B-15

EMULATING PHYSICAL CAPABILITIES

In addition to emulating devices of different screen sizes, you also have the option to emulate different hardware capabilities. When creating a new AVD, clicking the New . . . button will display a dialog for choosing the type of hardware you want to emulate (see Figure B-16).

For example, if you want to emulate an Android device with no touch screen, select the "Touch-screen support" property and click OK. Back in the AVD dialog, change the value of the property from yes to no (see Figure B-17).

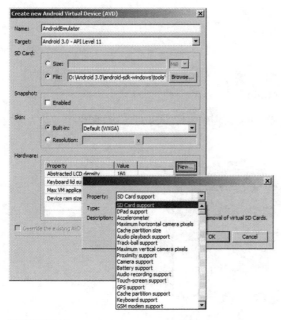

FIGURE B-16

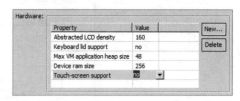

FIGURE B-17

This will create an AVD with no touch-screen support (i.e., users won't be able to use their mouse to click on the screen).

You can also simulate location data using the Android emulator. Chapter 4 discusses this in more detail.

KEYBOARD SHORTCUTS

The Android emulator supports several keyboard shortcuts that enable you to mimic the behavior of a real handset. The following list describes the shortcuts that you can use with the emulator:

➤ **Esc** — Back

➤ **Home** — Main screen

➤ **F2** — Toggles context-sensitive menu

➤ **F3** — Call Log

➤ **F4** — Lock

➤ **F5** — Search

➤ **F8** — Toggles data network (3G)

continues

(continued)

➤ **Ctrl+F5** — Ringer volume up

➤ **Ctrl+F6** — Ringer volume down

➤ **Ctrl+F11/Ctrl+F12** — Toggle orientation

For example, by pressing Ctrl+F11, you can change the orientation of the emulator to portrait mode (see Figure B-18).

FIGURE B-18

One useful tip to make your development more productive is to keep your Android emulator running during development — avoid closing and restarting it. Because the emulator takes time to boot up, it is much better to leave it running when you are debugging your applications.

SENDING SMS MESSAGES TO THE EMULATOR

You can emulate sending SMS messages to the Android emulator using either the Dalvik Debug Monitor Service (DDMS) tool available in Eclipse, or the Telnet client.

 NOTE *The Telnet client is not installed by default in Windows 7. To install it, type the following command line in the Windows command prompt:* `pkgmgr /iu:"TelnetClient"`.

Take a look at how this is done in Telnet. First, ensure that the Android emulator is running. In order to telnet to the emulator, you need to know the port number of the emulator. You can obtain this by looking at the title bar of the Android emulator window. It normally starts with 5554, with each subsequent emulator having a port number incremented by two, such as 5556, 5558, and so on. Assuming that you currently have one Android emulator running, you can telnet to it using the following command (replace 5554 with the actual number of your emulator):

```
C:\telnet localhost 5554
```

To send an SMS message to the emulator, use the following command:

```
sms send +651234567 Hello my friend!
```

The syntax of the `sms send` command is as follows:

```
sms send <phone_number> <message>
```

Figure B-19 shows the emulator receiving the sent SMS message.

FIGURE B-19

Besides using Telnet for sending SMS messages, you can also use the DDMS perspective in Eclipse. If the DDMS perspective is not visible within Eclipse, you can display it by clicking the Open Perspective button (highlighted in Figure B-20) and selecting Other.

FIGURE B-20

Select the DDMS perspective (see Figure B-21) and click OK.

Once the DDMS perspective is displayed, you will see the Devices tab (see Figure B-22), which shows the list of emulators currently running. Select the emulator instance to which you want to send the SMS message, and under the Emulator Control tab you will see the Telephony Actions section. In the Incoming number field, enter an arbitrary phone number and check the SMS radio button. Enter a message and click the Send button.

FIGURE B-21

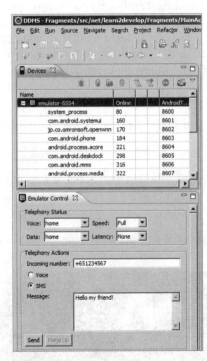

FIGURE B-22

The selected emulator will now receive the incoming SMS message.

If you have multiple AVDs running at the same time, you can send SMS messages between each AVD by using the port number of the emulator as the phone number. For example, if you have an emulator running on port number 5554 and another on 5556, their phone numbers will be 5554 and 5556, respectively.

MAKING PHONE CALLS

Besides sending SMS messages to the emulator, you can also use the Telnet client to make a phone call to the emulator. To do so, simply use the following commands.

 NOTE *At the time of writing, the Android 3.0 emulator does not support phone calls.*

To telnet to the emulator, use this command (replace 5554 with the actual number of your emulator):

```
C:\telnet localhost 5554
```

To make a phone call to the emulator, use this command:

```
gsm call +651234567
```

The syntax of the `gsm send` command is as follows:

```
gsm call <phone_number>
```

Figure B-23 shows the emulator receiving an incoming call.

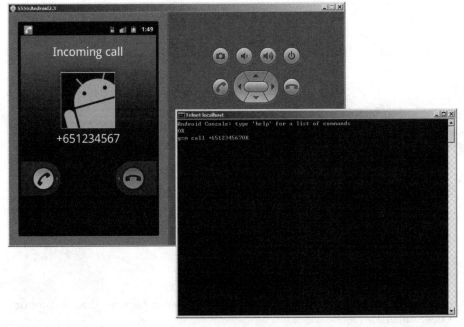

FIGURE B-23

Likewise, you can also use the DDMS perspective to make a phone call to the emulator. Figure B-24 shows how to make a phone call using the Telephony Actions section.

As with sending SMS, you can also make phone calls between AVDs by using their port numbers as phone numbers.

TRANSFERRING FILES INTO AND OUT OF THE EMULATOR

Occasionally, you may need to transfer files into or out of the emulator. The easiest way is to use the DDMS perspective. From the DDMS perspective, select the emulator (or device if you have a real Android device connected to your computer) and click the File Explorer tab to examine its file systems (see Figure B-25).

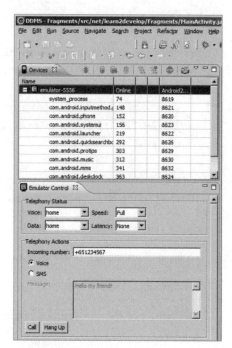

FIGURE B-24

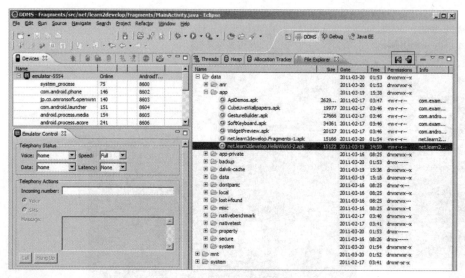

FIGURE B-25

The two buttons highlighted in Figure B-25 enable you to either pull a file from the emulator or push a file into the emulator.

Alternatively, you can also use the `adb.exe` utility shipped with the Android SDK to push or pull files to and from the emulator. This utility is located in the `<Android_SDK_Folder>`
`\platform-tools\` folder.

To copy a file from the connected emulator/device onto the computer, use the following command:

```
adb.exe pull /data/app/<filename> c:\
```

 NOTE *When using the* `adb.exe` *utility to pull or push files from or into the emulator, ensure that only one AVD is running.*

Figure B-26 shows how you can extract an APK file from the emulator and save it onto your computer.

```
C:\Windows\system32\cmd.exe                                    _ □ X

D:\Android 3.0\android-sdk-windows\platform-tools>adb.exe pull /data/app/net.lea
rn2develop.HelloWorld-2.apk c:\
68 KB/s (15122 bytes in 0.217s)

D:\Android 3.0\android-sdk-windows\platform-tools>_
```

FIGURE B-26

To copy a file into the connected emulator/device, use the following command:

```
adb.exe push NOTICE.txt /data/app
```

The preceding command copies the `NOTICE.txt` file located in the current directory and saves it onto the emulator's `/data/app` folder (see Figure B-27).

```
C:\Windows\system32\cmd.exe                                    _ □ X

D:\Android 3.0\android-sdk-windows\platform-tools>adb.exe push NOTICE.txt /data/
app
421 KB/s (10800 bytes in 0.025s)

D:\Android 3.0\android-sdk-windows\platform-tools>
```

FIGURE B-27

If you need to modify the permissions of the files in the emulator, you can use the `adb.exe` utility together with the shell option, like this:

```
adb.exe shell
```

Figure B-28 shows how you can change the permissions of the NOTICE.txt file by using the chmod command.

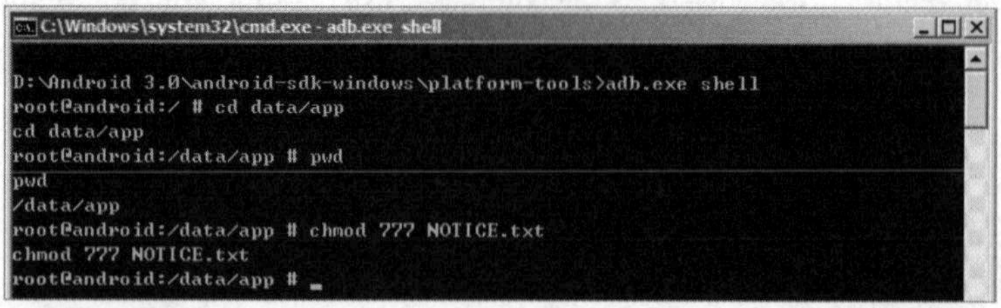

FIGURE B-28

Using the adb.exe utility, you can issue Unix commands against your Android emulator.

RESETTING THE EMULATOR

All applications and files that you have deployed to the Android emulator are stored in a file named userdata-qemu.img located in the C:\Users\<username>\.android\avd\<avd_name>.avd folder. For example, I have an AVD named AndroidTabletWithMaps; hence, the userdata-qemu.img file is located in the C:\Users\Wei-Meng Lee\.android\avd\AndroidTabletWithMaps.avd folder.

If you want to restore the emulator to its original state (to reset it, that is), simply delete the userdata-qemu.img file.

Answers to Exercises

This appendix contains the answers to the end of chapter exercises.

CHAPTER 1 ANSWERS

1. An AVD is an Android Virtual Device. It represents an Android emulator, which emulates a particular configuration of an actual Android device.

2. The `android:versionCode` attribute is used to programmatically check if an application can be upgraded. It should contain a running number (an updated application is set to a higher number than the older version). The `android:versionName` attribute is used mainly for displaying to the user. It is a string, such as "1.0.1."

3. The `strings.xml` file is used to store all string constants in your application. This enables you to easily localize your application by simply replacing the strings and then recompiling your application.

CHAPTER 2 ANSWERS

1. You can either use the `<fragment>` element in the XML file, or use the `FragmentManager` and `FragmentTransaction` classes to dynamically add/remove fragments from an activity.

2. One of the main differences between activities and fragments is that when an activity goes into the background, the activity is placed in the back stack. This allows an activity to be resumed when the user presses the Back button. Conversely, fragments are not automatically placed in the back stack when they go into the background.

3. Adding action items to an Action Bar is similar to creating menu items for an options menu — simply handle the `onCreateOptionsMenu()` and `onOptionsItemSelected()` events.

CHAPTER 3 ANSWERS

1. The dp unit is density independent and 160dp is equivalent to one inch. The px unit corresponds to an actual pixel on screen. You should always use the dp unit because it enables your activity to scale properly when run on devices of varying screen size.

2. With the advent of devices with different screen sizes, using the AbsoluteLayout makes it difficult for your application to have a consistent look and feel across devices.

3. For radio buttons, you need to use the setOnCheckedChangeListener() method on the RadioGroup to register a callback to be invoked when the checked RadioButton changes in this group. When a RadioButton is selected, the onCheckedChanged() method is fired. Within it, you locate individual RadioButton and then call their isChecked() method to determine which RadioButton is selected.

4. The three specialized fragments are ListFragment, DialogFragment, and PreferenceFragment.

CHAPTER 4 ANSWERS

1. The likely reasons are as follows:

➤ No Internet connection

➤ Incorrect placement of the <uses-library> element in the AndroidManifest.xml file

➤ Missing INTERNET permission in the AndroidManifest.xml file

2. Geocoding is the act of converting an address into its coordinates (latitude and longitude). Reverse geocoding converts a pair of location coordinates into an address.

3. The two providers are as follows:

➤ LocationManager.GPS_PROVIDER

➤ LocationManager.NETWORK_PROVIDER

4. The method is addProximityAlert().

CHAPTER 5 ANSWERS

1. You can either programmatically send an SMS message from within your Android application or invoke the built-in Messaging application to send it on your application's behalf.

2. The two permissions are SEND_SMS and RECEIVE_SMS.

3. The Broadcast receiver should fire a new intent to be received by the activity. The activity should implement another BroadcastReceiver to listen for this new intent.

4. The permission is INTERNET.

CHAPTER 6 ANSWERS

1. You specify the minimum Android version required using the `minSdkVersion` attribute in the `AndroidManifest.xml` file.

2. You can either use the `keytool.exe` utility from the Java SDK, or use Eclipse's Export feature to generate a certificate.

3. Go to the Settings application and select the Applications item. Check the "Unknown sources" item.

INDEX

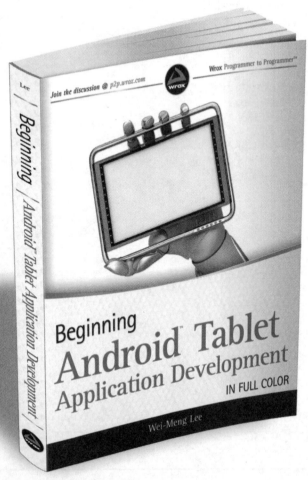